TRANSIT FACILITY GUIDE

JUNE 2026

ASIAN DEVELOPMENT BANK

6 ADB Avenue, Mandaluyong City, 1550 Metro Manila, Philippines
Tel +63 2 8632 4444; Fax +63 2 8636 2444
www.adb.org

ISBN 978-92-9277-758-6 (print); 978-92-9277-759-3 (PDF); 978-92-9277-760-9 (ebook)
Publication Stock No. TIM260135-2
DOI: http://dx.doi.org/10.22617/TIM260135-2

Cover design by Claudette Rodrigo.

CONTENTS

TABLES AND FIGURES

TABLES

FIGURES

ABBREVIATIONS

GEDSI	gender equality, disability, and social inclusion
m	meter
PUDO	pick-up and drop-off
TOD	transit-oriented development

EXECUTIVE SUMMARY

The *Transit Facility Guide* is designed to support urban transport stakeholders—including planners, designers, government officials, developers, and transit operators—in creating efficient, user-friendly, and well-integrated transit facilities. It is a condensed version of the Transit Facility Guideline developed for PT Mass Rapid Transit Jakarta (MRT Jakarta), a regional owned railways operator in DKI Jakarta Province, Indonesia, which was prepared with technical assistance from the Asian Development Bank (TA 9978-INO: Sustainable Infrastructure Assistance Program Phase II – Innovative Infrastructure Financing, Infrastructure Planning, and Program Management Support [Subproject 1]). TA 9978 is under the Sustainable Infrastructure Assistance Program II financed by the Government of Australia through its Department of Foreign Affairs and Trade.

This guide is prepared for brevity without reducing its functionality as a planning reference. It presents a clear, consistent, and actionable approach to planning and designing transit facilities, either in new station development or retrofits of existing facilities. The original guide was prepared for Jakarta City but offers scalable solutions that can be adapted to diverse local contexts. This guide aligns with urban sustainability, inclusive access, and climate resilience goals, and serves as a key technical reference for cities aiming for more livable and people-centered mobility.

The guide is structured around five core objectives or planning principles for people-centered transit facilities: enhancing multimodal connectivity, promoting people-centric local access, ensuring universal accessibility, fostering social and public well-being and safety, and integrating sustainability and future-proofing in design. These principles are supported by practical tools such as station and street typologies (Chapter 3), a scoring and evaluation framework (Chapter 5), and design elements by mode (Chapter 4).

By applying this guide, stakeholders can enhance user experience and accessibility, increase public transport usage, optimize land use around stations, and achieve long-term sustainability and resilience in urban transit systems.

1 INTRODUCTION

1.1 Background

As cities grow and mobility demands increase, the role of public transit in shaping sustainable, inclusive urban environments becomes ever more critical. Transit stations—particularly those serving urban rail and multimodal systems—are not just access points to transport, but dynamic public spaces that encourage and facilitate the use of public transport. A well-designed transit facility makes public transport a preferred choice for regular travel.

However, in many urban areas, the spaces surrounding transit stations remain underdeveloped and poorly integrated, lacking seamless multimodal connections, and inclusive design. These gaps in first- and last-mile infrastructure—such as pedestrian routes, convenient bicycle infrastructure, and well-organized pick-up and drop-off points—can undermine the effectiveness of public transport systems. They create barriers that limit accessibility for all users, reduce comfort and safety, and discourage people from choosing public transport as a viable option.

In response to these challenges, a clear and practical transit facility guide is essential for shaping transit facility spaces that work effectively for diverse users. Transit environments that prioritize safety, comfort, and inclusivity not only improve access and safety, but also overall user experience. They also transform transit stations into thriving urban hubs rather than mere points of transit.

1.2 Objective of the Guide

This guide provides a practical framework for planning and designing transit facilities around urban rail and multimodal stations. It supports cities in improving the physical integration between public transport and surrounding communities, ensuring transit environments are safe, inclusive, and easy to navigate.

The guide is structured around five core objectives or planning principles for people-centered transit facilities: enhancing multimodal connectivity, promoting people-centric local access, ensuring universal accessibility, fostering social and public well-being and safety, and integrating sustainability and future-proofing in design.

1.3 Scope and Applicability

This guide supports the planning and design of transit facilities in and around urban rail and multimodal stations. While grounded in experience from dense metropolitan areas in Jakarta City, its principles are adaptable to a wide range of settings, including emerging cities, secondary urban areas, and diverse station typologies. It applies to transit environments associated with systems such as mass rapid transit and light rapid transit; bus rapid transit and other high-capacity bus networks; commuter rail stations; and multimodal terminals with intermodal transfers.

The guide applies to both new transit developments—from early-stage planning to detailed design—and the retrofitting or upgrading of existing facilities, especially where accessibility, connectivity, or user experience needs improvement. It covers planning for first- and last-mile access and supporting infrastructure such as sidewalks, drop-off zones, bike parking, and bus stops, as well as wayfinding systems and intermodal transfer points. The focus is on public spaces and access elements in the unpaid zone (areas outside the ticketed platforms), while aligning with internal station designs where relevant.

The guide covers the following unpaid areas. These focus areas typically span a 50–400-meter radius from the station entrance, varying by station type, land use, and available space.

(i) Pedestrian facilities, such as sidewalks, crossings, ramps, and covered walkways
(ii) Bicycle infrastructure, including access routes and secure parking
(iii) Pick-up and drop-off areas for private vehicles, ride-hailing, and taxis
(iv) Bus stops and feeder vehicle bays, supporting seamless intermodal transfers
(v) Waiting areas, signage, and wayfinding systems, enhancing user orientation and comfort
(vi) Accessibility features, such as tactile paving, curb ramps, and universal design elements
(vii) Public space elements, such as landscaping, lighting, seating, and safety features that enhance the quality and usability of the area around the station

This guide does not provide technical specifications for internal station elements, such as fare gates, ticketing systems, or platform-level infrastructure. Those components are typically addressed in separate design guidelines or engineering standards issued by transit operators or infrastructure agencies. However, the planning of external transit facilities should be closely coordinated with internal station layouts to ensure smooth passenger flow and logical connections between modes.

1.4 Key Concepts and Terms

The following key concepts are used throughout the guide. Understanding these terms will enable readers to apply the design principles more effectively across different transit system and urban contexts.

Transit facilities. Physical elements that support access to and from a transit station, including sidewalks, crossings, waiting areas, pick-up and drop-off (PUDO) points, bicycle parking, signage, and related infrastructure. These facilities are typically located in the public realm outside the fare-paid zone.

(i) **First- and last-mile connectivity**. The portion of a passenger's journey that occurs before boarding and after alighting from a transit service. Effective design of first- and last-mile facilities ensures safe, efficient, and convenient access between stations and homes, workplaces, or other destinations.
(ii) **Unpaid zone.** Areas surrounding a transit station that are open to the public and do not require a ticket to access. These include entrances, sidewalks, access roads, intermodal transfer zones, and other shared spaces.
(iii) **Multimodal integration.** Coordination of different modes of transportation—such as rail, bus, walking, cycling, and ride-hailing—to provide a seamless journey for users. Integrated facilities reduce transfer times, improve accessibility, and encourage public transport use.
(iv) **Universal accessibility.** Design that ensures all users, including people with disabilities, older people, children, and those with limited mobility, can safely and comfortably access and use transit facilities.

(v) **Transit-oriented development (TOD).** A planning approach that promotes compact, mixed-use development within walking distance of high-capacity transit stations. TOD principles emphasize walkability, density, and multimodal connectivity.

1.5 How To Use The Guide

This guide is structured to support professionals and decision makers throughout all stages of transit facility planning and design. Users can read the full publication or refer to specific sections based on their project phase, role, or technical needs. Each chapter builds on a clear set of principles and offers actionable guidance, supported by visual references, mode-specific recommendations, and real-world examples.

The guide has the following key features:

(i) **Design principles**. Ensure that new transit facilities are systematically planned and integrated.

(ii) **Station and street typologies**. Guide on how to plan for different station types and surrounding street environments, tailored to varying levels of urban intensity and multimodal complexity.

(iii) **Transit facility elements.** Detailed guide by mode (pedestrian, bicycle, bus, rail, private vehicles, ride-hailing) to support seamless and safe transfers at station areas.

(iv) **Evaluation framework.** A simplified tool to help planners and agencies assess the quality and effectiveness of transit facilities and prioritize improvements.

(v) **Case examples.** Short, illustrative examples demonstrating how the guide can be applied in real-world station area contexts.

This flexible structure enables the guide to serve multiple purposes: as a reference manual during planning, design, or review processes; as a training or orientation tool for project teams and agency staff; or as a framework to support policy alignment, design consistency, and project scoping. Whether planning a new transit corridor, upgrading an existing station, or developing a city-wide multimodal strategy, users can adapt the guide to fit local conditions, priorities, and institutional capacity.

1.6 Structure of the Guide

This guide is organized into five chapters, each building on the principles and offering practical insights for transit facility planning and design. Its flexible structure allows users to read the entire publication or refer to relevant sections depending on their role, project phase, or technical needs.

Chapter 1 – Introduction provides the background, objectives, scope and applicability, key concepts and terms, how to use the guide, and the structure of the guide. **Chapter 2 – Planning Principles** outlines the foundational design principles for inclusive, accessible, and connected station areas, including multimodal integration; gender equality, disability, and social inclusion (GEDSI); wayfinding; and sustainability. **Chapter 3 – Station Typologies and Street Contexts** introduces classification systems for different types of stations and surrounding street environments to support context-sensitive planning. **Chapter 4 – Transit Facility Elements by Mode** provides a detailed guide on key access elements associated with different transport modes—pedestrian, bicycle, bus, rail, private vehicles, and ride-hailing—and how they should be integrated within station areas. **Chapter 5 – Evaluation and Scoring Framework** presents a simplified tool to assess the completeness and quality of transit facility design and identify opportunities for improvement.

PLANNING PRINCIPLES 2

High-quality transit facilities go beyond functionality, serving as public spaces that shape how people move, interact, and experience the city. Planning and designing must extend beyond technical specifications, drawing on principles that reflect social, spatial, and environmental priorities.

The core objectives or planning principles that guide the development of transit facilities around urban rail and multimodal stations are as follows:

(i) enhancing multimodal connectivity,

(ii) promoting people-centric local access,

(iii) ensuring universal accessibility,

(iv) fostering social and public well-being and safety, and

(v) integrating sustainability and future-proofing in design.

This chapter presents these core objectives. These principles apply across various city contexts, station typologies, and transport modes, reflecting the shared goal of creating accessible, safe, inclusive, and resilient transit environments that support broader urban development objectives.

The principles are examined around four key areas:

(i) core design objectives for transit facilities;

(ii) GEDSI;

(iii) wayfinding and visual legibility to support intuitive passenger movement; and

(iv) environmental sustainability and climate resilience in infrastructure design.

The principles jointly form a foundation for making design decisions that are user-focused, context-sensitive, and future-ready.

2.1 Core Design Objectives for Transit Facilities

The success of a transit system depends not only on routes, frequency, or vehicles, but also on the quality of access to stations. Well-designed transit facilities enable passengers to safely and comfortably reach, enter, and transfer between modes.

This section outlines the five core objectives that define a comprehensive, people-centered transit experience:

(i) enhancing multimodal connectivity,

(ii) promoting people-centric local access,

(iii) ensuring universal accessibility,
(iv) fostering social and public well-being and safety, and
(v) integrating sustainability and future-proofing in design.

2.1.1 Enhancing Multimodal Connectivity

Goal: Enable seamless and user-friendly transfers between modes (rail, bus, bicycle, walking, and ride-hailing) to ensure a smooth, cohesive, and time-efficient journey.

What to consider in practice: The physical boundary of interest typically includes the zone within a 50–400-meter (m) radius from the station entrance, depending on the station type, surrounding land use, and available space.

(i) Minimize transfer distances and walking times between modes
(ii) Locate intermodal facilities within close proximity to station entries
(iii) Provide continuous, safe, and barrier-free connections between facilities
(iv) Integrate schedules and signage where feasible to enhance clarity and ease of navigation

Design tip: Use visual alignment. Align physical entrances and circulation paths so users can easily see the next mode without relying solely on signage.

Common pitfalls to avoid:

(i) Disconnected modes with poor signage or indirect transfer routes
(ii) Conflicts between walking paths and parking areas
(iii) Long or unsafe walking distances between bus and rail platforms

2.1.2 Promoting People-Centric Local Access

Goal: Prioritize walking and cycling for local station access to enhance safety, convenience, and user satisfaction, especially for short and frequent trips.

What to consider in practice:

(i) Provide sidewalks with a 2.5–4.0 m minimum clear width, depending on expected pedestrian volume
(ii) Include buffer zones such as bollards or trees to separate pedestrians from motorized traffic
(iii) Implement traffic calming measures such as raised crossings and narrow entry points around station perimeters to enhance safety

Checklist for people-centric access:

(i) Continuous, nonslip sidewalks
(ii) Covered walkways or trees for shade
(iii) Clearly marked pedestrian crossings
(iv) Visual cues guiding movement to/from station

People-centric access and entrance. (left photo) A station in Taipei,China provides an accessible environment with wide, continuous sidewalks, physical separation from traffic, and covered pedestrian access; (right photo) A covered station entrance provides direct, legible, and comfortable pedestrian access (photos by rubensukatendel in unsplash).

2.1.3 Ensuring Universal Accessibility

Goal: Design spaces that can be used independently by people of all ages, abilities, and backgrounds.

What to consider in practice:

(i) Ensure step-free access from public realm to station entrances

(ii) Provide audible signals, tactile paving, and high contrast in signage

(iii) Add seating and rest points on long or steep access routes

Design tip: Design for those with the greatest needs, and everyone benefits. Wide paths for wheelchairs also support strollers and luggage.

Common pitfalls to avoid:

(i) Missing curb ramps or misaligned crossings

(ii) Tactile blocks directing users into hazards

(iii) Narrow entry points or steep slopes without resting areas

Universal accessibility. (left photo) A step-free ramp with adequate width, handrails, and gentle gradients enables independent and safe access for wheelchair users and others with mobility needs; (right photo) A station entrance should have step-free ramp access, adequate width, and handrails (photos by Made for Movement and Myanya Khintan Sabrins).

2.1.4 Fostering Social and Public Well-Being and Safety

Goal: Ensure station surroundings are safe, comfortable, and welcoming for all users, even at night or during off-peak hours.

What to consider in practice:

(i) Design for natural surveillance with open sightlines and no hidden corners

(ii) Ensure good lighting at access points, walkways, and waiting zones

(iii) Provide seating, shaded areas, and drinking water where possible

(iv) Activate public spaces around the station to encourage social interaction

Good design elements:

(i) Transparent shelters

(ii) Lighting that minimizes shadows

(iii) Entry zones facing active streets or open plazas

(iv) Seating and greenery near transit stations

Public well-being and safety around stations. (left photo) Lighting provision in the Netherlands; (right photo) drinking fountain in a Sydney metro station (photos by Lightronics, Netherlands, and Civiq).

2.1.5 Integrating Sustainability and Future-Proofing in Design

Goal: Plan facilities that are eco-friendly, resource-efficient, and adaptable to evolving transport needs and climate conditions.

What to consider in practice:

(i) Use permeable paving and drainage in PUDO zones

(ii) Integrate trees, bioswales, or green buffers in the design

(iii) Design drop-off and waiting areas adaptable for new modes (e.g., e-scooters, bike-sharing)

Case Example

Yogyakarta Bus Rapid Transit Stops

Yogyakarta's bus rapid transit stops use solar-powered lighting systems to cut electricity costs and improve passenger safety. This renewable energy system ensures continuous lighting during outages, offering a low-cost, climate-resilient solution that enhances service reliability—especially in areas with frequent power disruptions or limited grid access.

Checklist: Sustainability in design

(i) Shade trees or green buffers

(ii) Passive cooling (e.g., airflow corridors)

(iii) Climate-appropriate, low-maintenance materials

(iv) Modular layout for future mode integration to and from the station

2.2 Gender Equality, Disability, and Social Inclusion

Inclusive transit facilities are critical to ensure public transport serves everyone fairly regardless of gender, age, disability, socioeconomic status, or other factors. The physical and social design of spaces around transit stations such as sidewalks, waiting areas, pedestrian crossings, and intermodal zones directly affects how people feel safe, welcome, and able to use public transport. Applying a GEDSI approach to transit facility planning means going beyond basic accessibility standards. It involves actively understanding and addressing the diverse needs of users, especially those most vulnerable to exclusion or harm in public spaces.

2.2.1 Understanding Intersectional Barriers

Transit facilities should consider the needs of the following groups:

(i) Women and girls who often prioritize personal safety, clear sightlines, gender-responsive facilities, and good lighting, particularly at night

(ii) People with disabilities (physical, sensory, intellectual, psychosocial) who need step-free access, tactile and auditory cues, consistent curb heights, accessible customer service, easy-to-understand signage, quiet and low-stimulation spaces, simple layouts, clear wayfinding, options for assisted travel, and staff trained in mental health awareness to ensure a supportive and inclusive environment

(iii) Older people who need rest areas, gentle slopes, clear wayfinding, and good visibility

(iv) Pregnant women and users with temporary impairments, who require wider paths, resting spots, and barrier-free surfaces

(v) Children and caregivers who need wider pathways, stroller-friendly access, safe crossing points, and designated waiting zones

(vi) Low-income communities, which depend on affordable, informal, or active transport (walking, cycling, or motorcycle taxis) for first- and last-mile connectivity, and are most affected by fare- and distance-based pricing

Acknowledging these diverse needs enables planners to design inclusive transit environments, avoiding exclusion caused by one-size-fits all approaches. Inclusive transit design ensures access and safety are standard rights for everyone, not privileges.

2.2.2 Key Design Strategies for Gender Equality, Disability, and Social Inclusion Integration

Inclusive access requires integrated design covering safety, security, comfort, affordability, and easy multimodal wayfinding. Step-free access with ramps, elevators, gentle slopes, and nonslip, barrier-free pathways must connect public spaces to station entrances.

Tactile paving and audible cues are essential for persons with visual impairments. Level, unobstructed surfaces support wheelchair users, caregivers, and older people. Public areas should be open, well-lit, and free of hidden corners with clear sightlines to enhance safety especially for women, children, and older people. Transparent shelters, well-placed seating, and active frontage improve visibility and comfort.

Facilities should also offer shaded seating, stroller-friendly pathways, and gender-sensitive amenities, including accessible toilets, nursing rooms, family waiting areas, and quiet spaces for neurodiverse users.

Clear, inclusive communication is essential. Wayfinding should include multilingual signs with intuitive symbols, high contrast, readable fonts, and audio announcements to support diverse users, including tourists, people with disabilities, and non-native speakers.

Engaging women, people with disabilities, older people, and local communities in the planning process is essential. Walkability audits, accessibility testing, and participatory design reviews guide decisions, supported by feedback for continuous improvement.

These strategies are especially critical for low-income groups, who rely on walking, informal transport, or bicycles for first- and last-mile connectivity and are most sensitive to fare changes or displacement. Prioritizing their needs contributes to a more equitable, resilient, and inclusive transit system for all.

2.3 Wayfinding and Visual Legibility to Support Intuitive Passenger Movement

A well-designed transit area helps passengers navigate intuitively and confidently with minimal confusion. Wayfinding goes beyond signage, it combines visual cues, clear circulation paths, and consistent information. Effective wayfinding improves comfort, reduces anxiety, improves perceptions of safety, and increases overall system efficiency. It is especially critical for first-time riders, tourists, older people, and people with disabilities who rely on clarity and predictability to travel independently and confidently. The key principles that a good wayfinding strategy should consist of are presented in Figure 1, which shows the interactive process between the individual and the environment for wayfinding.

Good wayfinding design helps pedestrians select safe, direct routes to transit access points or destinations. It supports users in identifying transfer opportunities, finding exits, and locating services within and around the station.

Effective signage guides passengers from any starting point and supports spatial orientation throughout the trip. Good wayfinding systems enhance navigation and create a more inclusive and efficient transit experience for all.

Figure 1: What is Wayfinding

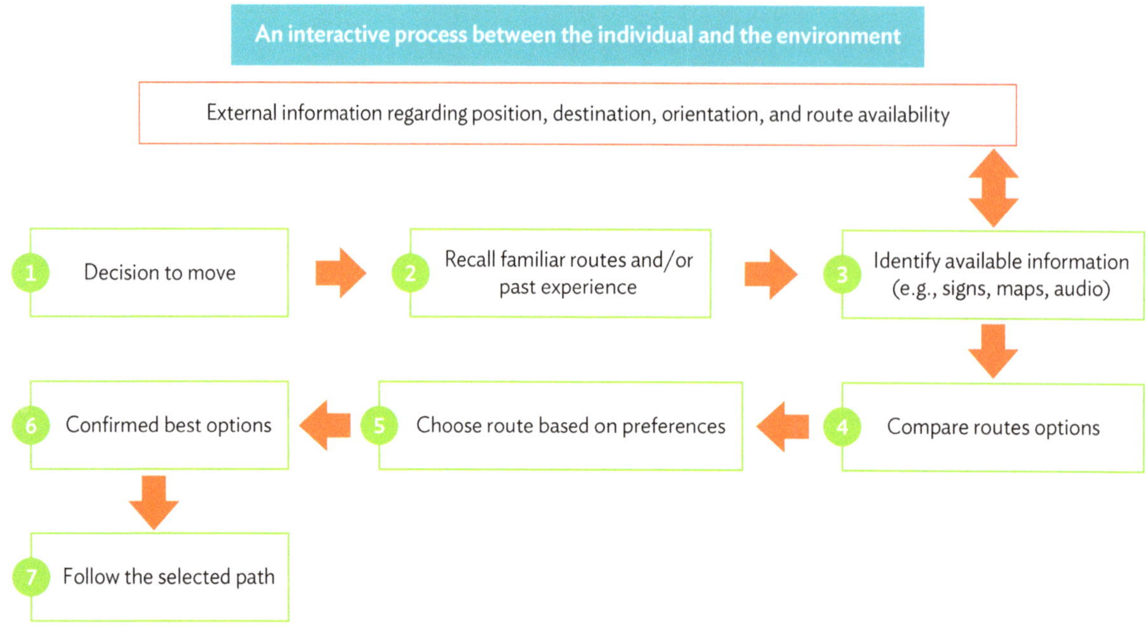

Source: Asian Development Bank.

Figure 2: Wayfinding Objectives and Principles

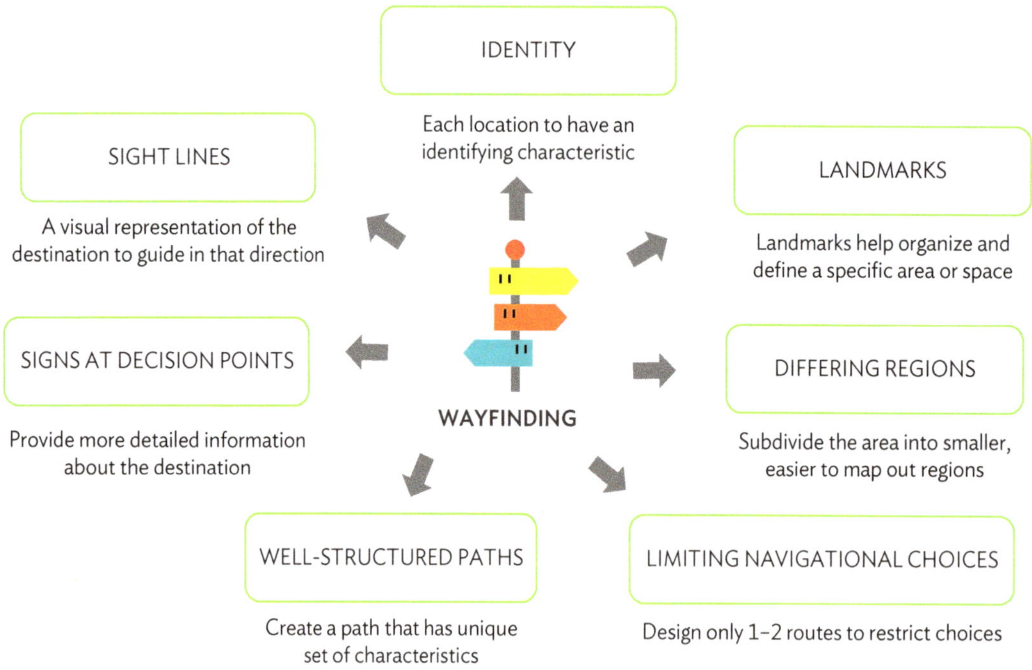

Source: Asian Development Bank.

2.3.1 Key Principles of Wayfinding Design

Effective wayfinding relies on clarity, consistency, and spatial logic. Information should be simple, concise, and free of visual clutter, allowing users to easily identify their current location and determine their next steps. A consistent visual language, using color codes, icons, maps, and fonts should be applied across all transit modes and placed at key decision points such as entrances, intersections, and transfer areas. Legible, open spaces with minimal visual obstructions reduce the need for excessive signage. Architectural features such as pathway alignment, lighting direction, floor patterns, and ceiling lines can also guide movement and support intuitive navigation.

Wayfinding should be inclusive and multisensory, combining various communication formats such as signage, pavement markings, lighting, digital displays, and tactile guidance. This supports users with visual, hearing, mobility, or language challenges. Messages should be clear across multiple formats, e.g., multilingual texts, universal icons, audible announcements, and accessible signage to help all users navigate transit areas independently and confidently.

Seamlessly integrated wayfinding helps passengers navigate stations smoothly and confidently, even in unfamiliar or crowded spaces. It is not just decoration but a vital part of creating safe, inclusive, and user-friendly transit systems.

Table 1: Practical Wayfinding Elements in Transit Facilities

Element	Role and Inclusive Features
Entry Signage	Clearly identifies the station name and primary access points from a distance. Select high-contrast fonts (minimum 70% contrast ratio) and large icons to identify station names and entrances from a distance. Include tactile letters and Braille panels at reachable heights.
Directional Signage	Guides users to platforms, exits, intermodal transfers, and key amenities. Use color-coded routes (e.g., blue for platforms, green for exits) and clear restroom signage. Position signs at both eye level (1.2–1.5 m) and knee level (0.5–0.8 m) for wheelchair users and children.
System Maps	Provides an overview of the transit network, routes, and major destinations. Provide simplified, large-print schematic maps with color distinctions for each line. Place at mid-height (1.0 m) for wheelchair access and include audio-playback buttons for persons with low vision.
Real-time Information Displays	Displays up-to-date information on arrivals, delays, and service changes. Use clear, sans serif fonts at a minimum of 30 pts for easy reading. Add audio announcements for arrival and departure updates. Include contrast-adjustable screens for low-light conditions.
Pavement Markings	Reinforces direction and defines safe walking routes, especially in shared zones. Install high-visibility, nonslip ground arrows and tactile directional tiles leading to platforms, elevators, and exits. Ensure that markings extend into shared zones with consistent color standards.
Landmarks and Visual Anchors	Aids spatial orientation, particularly in large, complex, or multilevel stations. Integrate gender-neutral artwork or community sculptures at major intersections to serve as reference points. Include lighting accents to enhance visibility for after-dark travel, prioritizing women's safety.

m = meter.

Source: Asian Development Bank.

2.4 Environmental Sustainability and Climate Resilience in Infrastructure Design

Transit facilities are not only mobility infrastructure; they are also urban spaces with environmental impacts and long-term exposure to climate-related risks. Designing them for sustainability and resilience helps cities reduce emissions; manage natural resources more effectively; and adapt to climate challenges such as extreme heat, flooding, or sea-level rise.

This section outlines strategies for integrating environmental goals into the planning and design of transit facilities around urban rail and multimodal stations.

2.4.1 Sustainability in Transit Facility Design

Sustainable design enhances long-term environmental performance while improving user comfort and operational efficiency. Strategies include the use of low-carbon material, such as durable materials with low embodied carbon, energy-efficient lighting and ventilation, and integration of solar panels or off-grid lighting systems where feasible. Green and shaded spaces including tree planting, vegetation buffers, and bioswales reduce temperatures, improve air quality, and enhance passenger comfort. Effective stormwater management, with features like permeable surfaces, proper drainage systems, and green infrastructure prevent flooding and water pooling near access points.

2.4.2 Climate Resilience in Station Area

Transit infrastructure must be designed to withstand and recover from environmental shocks, while ensuring safe access and evacuation for all users, including vulnerable groups, during and after adverse events. Resilient design must prioritize inclusive service continuity and equitable protection during extreme weather and emergency.

Key principles include the following:

Risk-aware siting

(i) Avoid placing transit infrastructure in flood-prone zones without adequate mitigation.

(ii) Elevate universally accessible paths (e.g., step-free ramps, tactile-guided routes) and critical facilities in flood or coastal risk zone.

Material and system resilience

(i) Use materials resistant to heat, moisture, and temperature change.

(ii) Ensure wayfinding and communication systems are weatherproof; multisensory (visual, tactile, and auditory); and accessible to all users.

Redundancy and flexibility

(i) Provide multiple, universal access routes and alternative circulation paths that are usable by wheelchair users, persons with vision or hearing impairments, pregnant women, and caregivers with strollers in case of partial disruptions.

(ii) Design public spaces and waiting zones that can be repurposed for inclusive evacuation or emergency shelter use with adequate lighting; visibility; and amenities such as safe spaces for women and children, accessible toilets, etc.

Design references for climate change adaptation. (upper left) Terraced green roofs and vertical gardens to absorb stormwater, reduce urban heat island effects, and improve thermal comfort around stations; (upper right) shaded façade with high-performance glazing reduces solar heat gain and cooling loads, enhancing energy resilience during heat waves; (lower left) shaded park with misting for passive cooling, shade trees, wind corridors, and low-water misting, which lower ambient temperatures for safer, cooler waiting areas; (lower right) blue-green corridor integrated bioswales and retention channels manage floods, recharge groundwater, and create climate-robust public realms (photos by WOHA [upper left]; tecnocooling [upper right and lower left]; lewincfl [lower right]).

STATION TYPOLOGIES AND STREET CONTEXTS 3

Transit stations are integral parts of a larger urban fabric, influencing how people access, experience, and interact with public transport. Designing effective transit facilities requires understanding the station type, surrounding street network characteristics, and access zones to support multimodal integration.

This chapter introduces three key classifications:

(i) station typologies based on physical form and operational function;

(ii) street contexts based on road function, traffic intensity, and urban form; and

(iii) access zones that define how the space around the station is organized for different users and modes.

These distinctions are especially useful during project inception or early design review phases, when high-level decisions about station form and street integration need to be made quickly and clearly—areas where many teams often face challenges.

3.1 Station Typologies

Transit station design depends on its form, location, and role within the transport network. Identifying the station typology is essential to choosing the right access facilities, spatial configurations, and public realm strategies. Each typology presents different opportunities and constraints for pedestrian flow, multimodal integration, and land use interaction.

This section outlines four common station typologies for urban rail and multimodal systems. While not exhaustive, they offer a foundation for context-sensitive planning and design. Figure 3 shows the Mass Rapid Transit station typology classification based on the MRT Jakarta case.

3.1.1 Underground Station

Definition: Station platforms are fully below street level, with access via stairs, ramps, escalators, or elevators.

Design implications:

(i) Requires vertical circulation design (elevators, escalators, and stairways)

(ii) Smaller footprint at street level allows more flexible public space design

(iii) Needs clear surface-level wayfinding and multiple entry points

Common use cases: Dense central business districts or heritage-sensitive areas.

As illustrated in Figure 4, a typical underground station layout organizes passenger circulation vertically while minimizing surface disruption.

Figure 3: Mass Rapid Transit Station Typology Classification (MRT Jakarta Case)

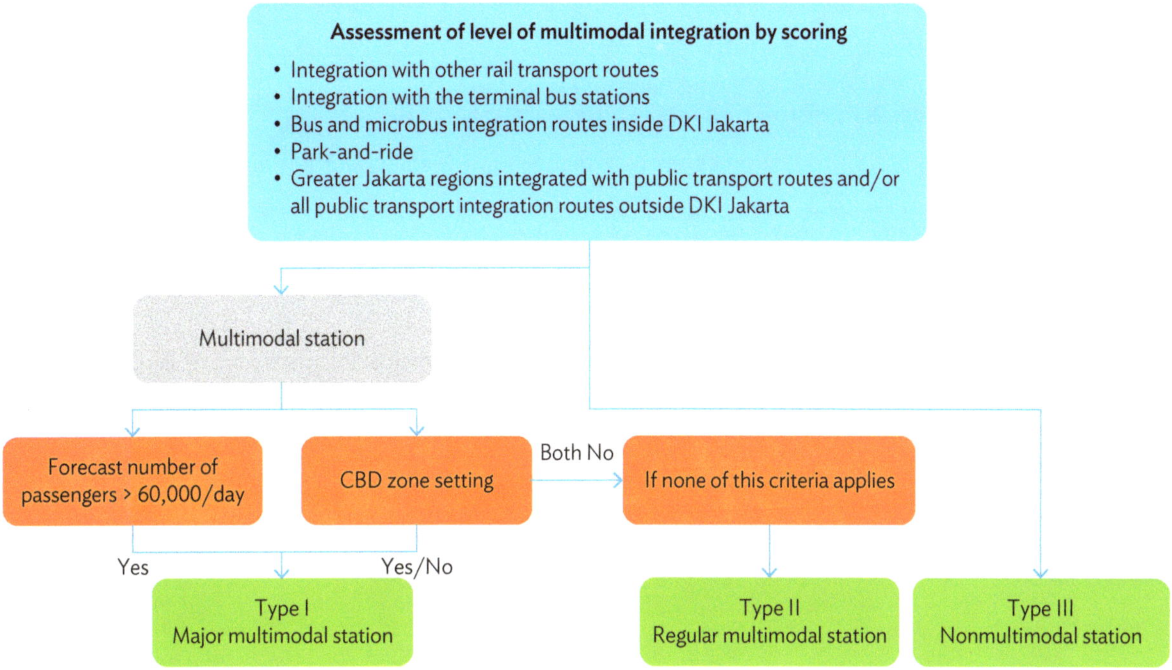

CBD = Central Business District.

Source: Asian Development Bank.

Figure 4: Typical Underground Station Layout

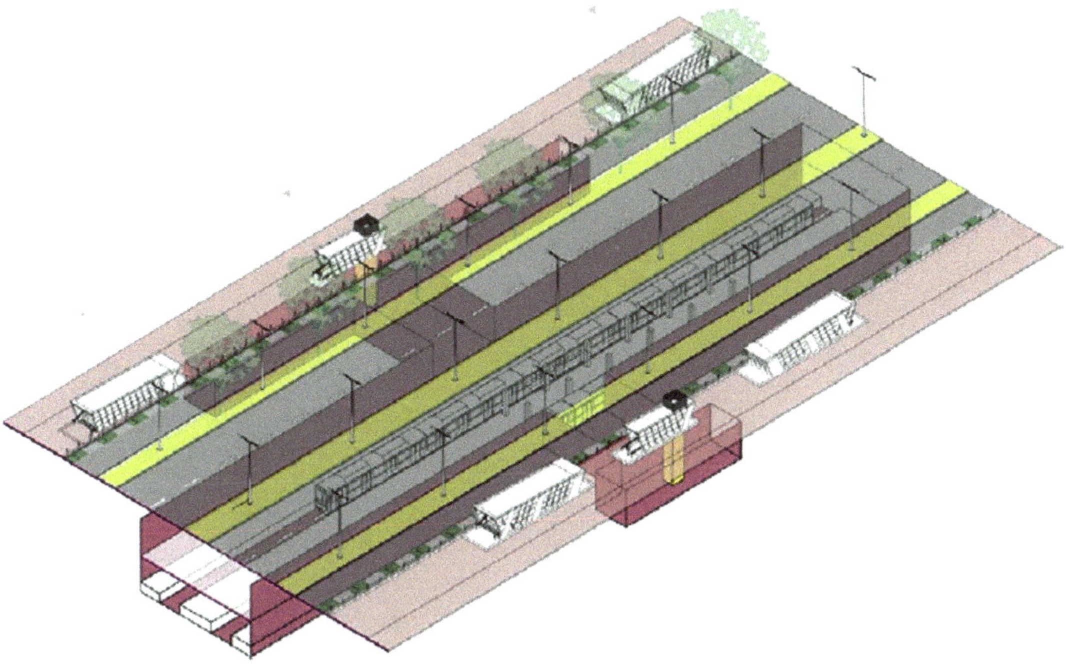

Source: Asian Development Bank.

3.1.2 Elevated Station

Definition: Station platforms are raised above street level, typically on viaducts or bridges.

Design implications:

(i) Requires vertical access infrastructure (stairs, lifts, escalators)

(ii) Must consider how piers and station columns affect ground-level movement

(iii) Public realm design should accommodate shaded space and active uses below the structure

Common use cases: Arterial corridors, outer urban areas, or flood-prone zones.

As shown in Figure 5, a typical elevated station layout emphasizes vertical connectivity while integrating the supporting structure with ground level.

Figure 5: Typical Elevated Station Layout

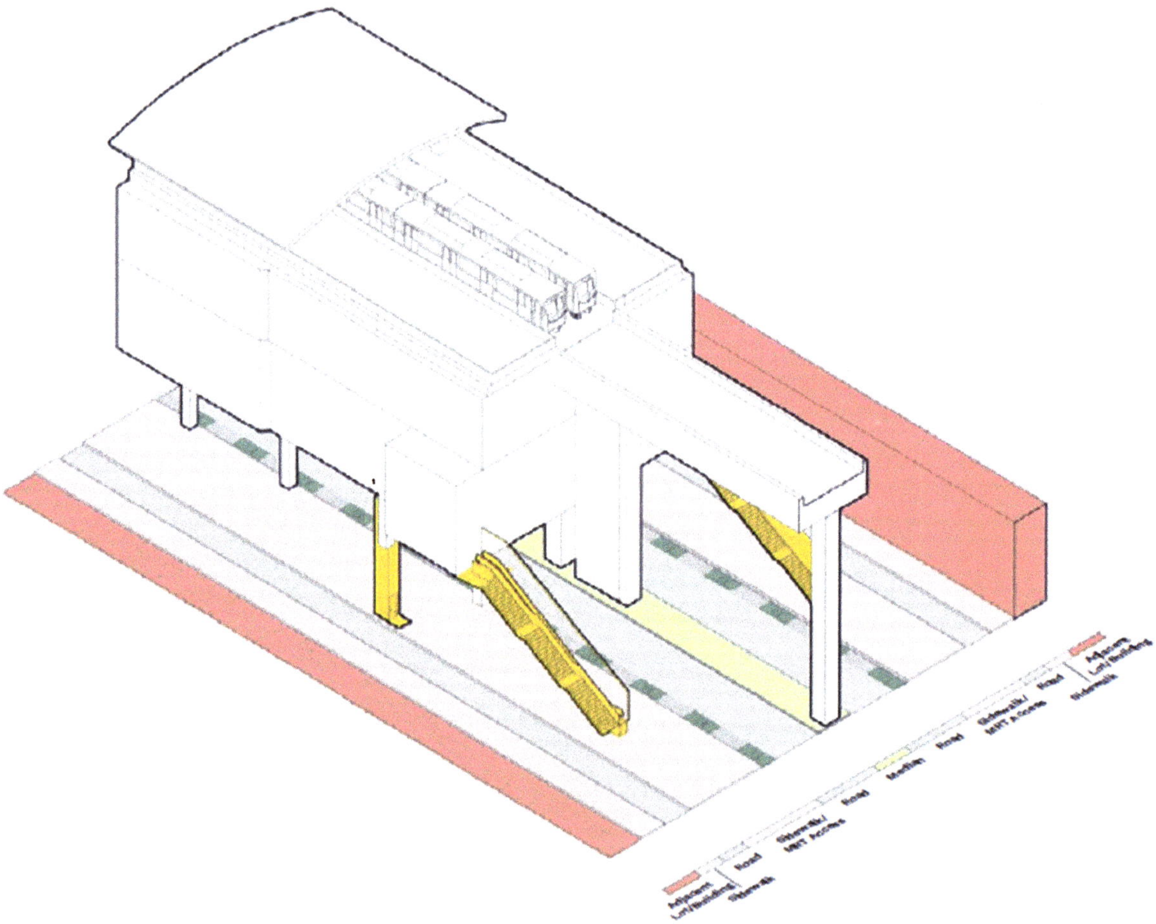

Source: Asian Development Bank.

3.1.3 At-Grade Station

Definition: Station platforms are located at the same level as the surrounding street or railway.

Design implications:

(i) Allows for easier access with minimal vertical infrastructure

(ii) Requires strong pedestrian safety measures at crossings and approach paths

(iii) Offers opportunities for strong integration with adjacent land uses

Common use cases: Suburban corridors, medium-density urban areas, or where cost efficiency is critical.

3.1.4 Terminal or Interchange Station

Definition: A station that functions as a major transfer point or end-of-line stop for multiple modes.

Design implications:

(i) Must accommodate high passenger volumes and multiple access routes

(ii) Requires spatial separation and coordination between modes (e.g., buses, paratransit, and bicycles)

(iii) Often integrated with retail, offices, or public services

Common use cases: City gateways, regional hubs, TOD centers.

Each station typology influences how transit facilities are placed, sized, and connected to the surrounding urban environment. Understanding these differences ensures design decisions are tailored to operational needs, space availability, and user behavior.

3.2 Street Context Classifications

The effectiveness of a transit station is strongly influenced by the type of street it faces. A well-integrated station responds not only to transit needs but also to the surrounding street's function, scale, and design characteristics. Understanding these street contexts helps planners position access points, manage traffic interactions, and design safe, inclusive public spaces.

This section introduces four simplified street context classifications commonly found around urban rail or multimodal stations. Each has distinct implications for pedestrian movement, multimodal access, and the public realm.

3.2.1 Primary Arterial Street

Definition: A wide, high-traffic road serving long-distance vehicle movement, often with multiple lanes in each direction.

Characteristics:

(i) High vehicle speeds and volumes

(ii) Limited pedestrian crossings and wider building setbacks

(iii) Frequent presence of flyovers, intersections, or commercial frontages

Design considerations:

(i) Provide grade-separated or signal-controlled pedestrian crossings

(ii) Create safe buffer zones between sidewalks and roadways

(iii) Design drop-off areas that do not interrupt through-traffic

As illustrated in Figure 6, a typical primary arterial street cross-section prioritizes vehicular flow while incorporating pedestrian safety measures through spatial separation.

Figure 6: Typical Primary Arterial Cross-Section

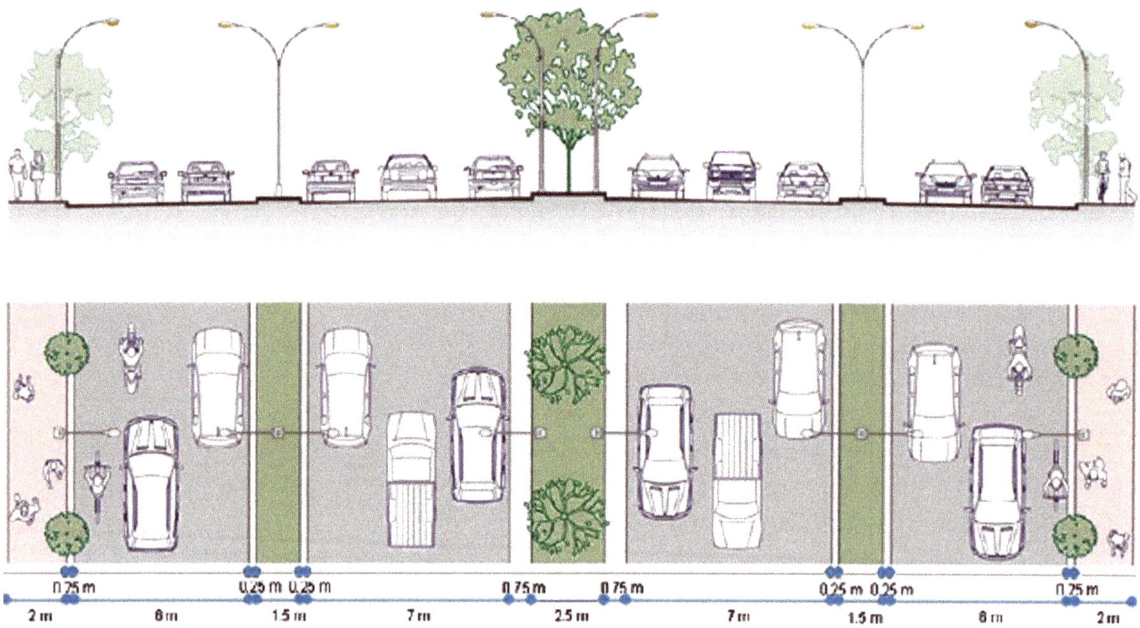

Source: Asian Development Bank.

3.2.2 Secondary or Collector Street

Definition: A medium-width road connecting local streets to arterials, often with moderate traffic and mixed-use frontages.

Characteristics:

(i) Balanced vehicle and pedestrian use

(ii) On-street parking, occasional bus routes, and local businesses

(iii) Moderate crossing distances and more flexible curb space

Design considerations:

(i) Enable shared spaces between modes

(ii) Include bicycle lanes or traffic calming near station zones

(iii) Allow for short-term parking or flexible use zones near station access points

As shown in Figure 7, a typical collector street cross-section supports multimodal movement while maintaining a pedestrian-friendly street environment.

Figure 7: Typical Collector Road Cross-Section

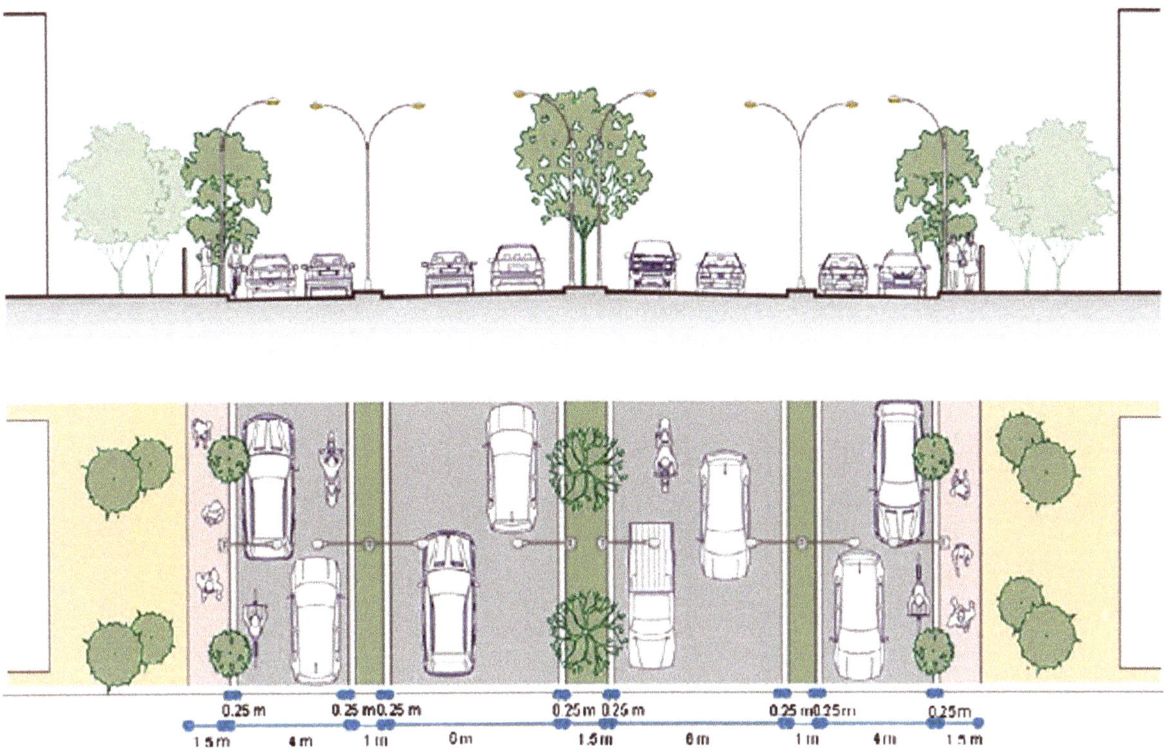

Source: Asian Development Bank.

3.2.3 Local or Neighborhood Street

Definition: A narrow, low-traffic road serving primarily residential or small-scale commercial buildings.

Characteristics:

(i) Slower speeds, lower vehicle volume

(ii) Higher walkability and community interaction

(iii) Potential for shared use by pedestrians, cyclists, and cars

Design considerations:

(i) Prioritize pedestrian paths and informal gathering spaces

(ii) Ensure clear directional signage to the station

(iii) Design for low-impact PUDO with minimal infrastructure

As illustrated in Figure 8, a typical local road cross-section fosters a pedestrian-friendly and community-oriented street environment.

Figure 8: Typical Local Road Cross-Section

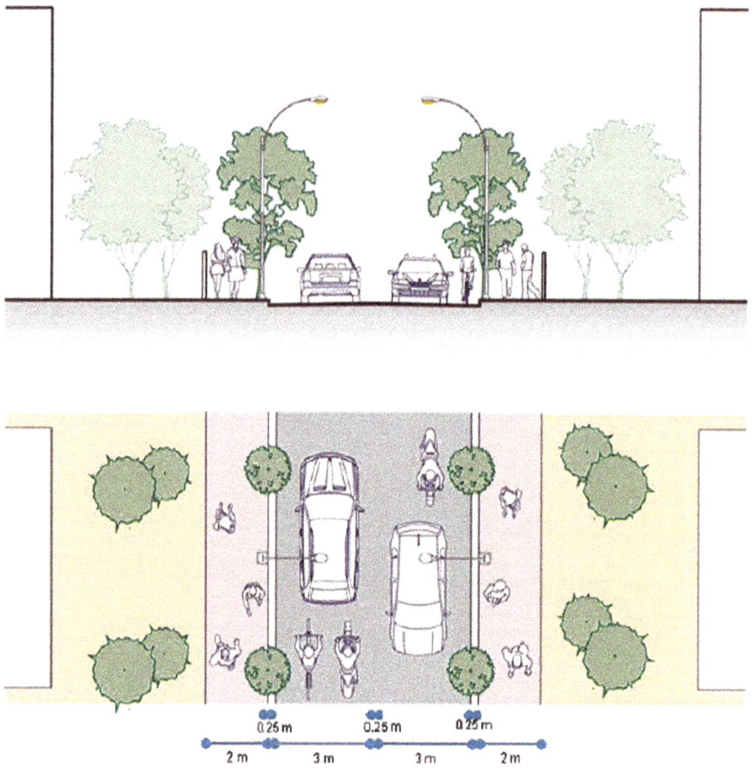

Source: Asian Development Bank.

3.2.4 Pedestrian Priority Street or Shared Space

Definition: A space where pedestrian movement dominates, often fully or partially car-free, sometimes shared with bicycles or emergency access vehicles.

Characteristics:

(i) High foot traffic, vibrant public realm, often located in TODs or civic spaces

(ii) Plazas, shaded walkways, retail frontage, seating areas

Design considerations:

(i) Maximize wayfinding and visibility of station access

(ii) Integrate seating, lighting, greenery, and universal design

(iii) Ensure emergency access and service vehicle routing without compromising walkability.

Understanding the street context helps determine feasible design features, public space allocation, and necessary safety measures around the station. It also supports coordination between transit agencies and street management authorities.

From a user's perspective, it can be the difference between a pleasant 3-minute walk to the station or a stressful, unsafe crossing through high-speed traffic. This is where street context directly affects ridership experience.

3.3 Access Zones and Hierarchies

The transit station areas serve multiple functions, including passenger access, vehicle movement, drop-off, public gathering, and sometimes commercial use. To ensure safe and efficient circulation, these areas should be organized into clear access zones with specific roles and design priorities.

This section presents a simplified zoning hierarchy to guide the layout of transit facilities and user circulation around station entrances. These zones are functional designations rather than fixed boundaries, helping to allocate space in a logical and user-friendly manner.

As shown in Figure 9, an example of mass rapid transit station coverage in MRT Jakarta's Dukuh Atas Station illustrates how access zones can be organized to optimize circulation and station functionality.

Figure 9: Example of Mass Rapid Transit Station Coverage Area (MRT Jakarta Dukuh Atas Station)

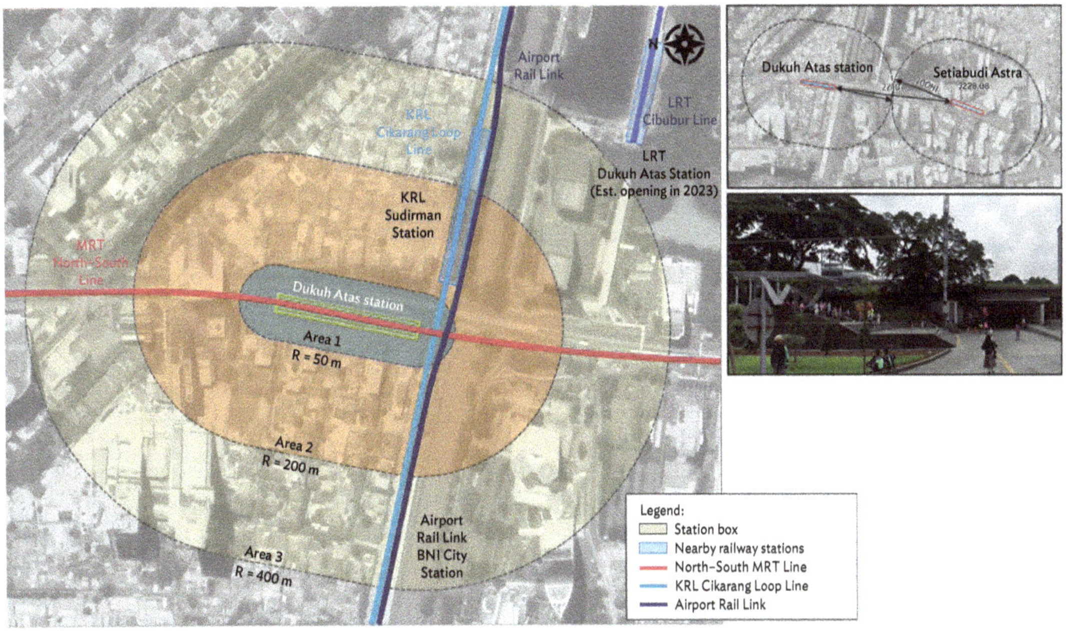

Source: Asian Development Bank.

3.3.1 Primary Access Zone

Purpose: Direct access for pedestrians entering or exiting the station

Location: Immediately adjacent to station entrances (usually within 0–15 meters)

Key features:

(i) Wide, obstruction-free sidewalks

(ii) Seating, lighting, wayfinding signage

(iii) Step-free routes and tactile paving

Design priority: Highest—prioritize pedestrian movement and comfort

3.3.2 Secondary Access Zone

Purpose: Facilitates connections with other modes and circulation

Location: Between 15 and 50 meters from station entrances

Key features:

(i) Bus bays, bike parking, ride-hailing PUDO zones

(ii) Taxi stands and feeder vehicle waiting areas

(iii) Shelter and designated waiting areas

Design priority: High—accommodate multimodal interchange, avoid pedestrian-vehicle conflicts

3.3.3 Tertiary or Supporting Zone

Purpose: Supports ancillary functions and transition to surrounding streets or developments

Location: Typically, 50–150 meters from the station entrance

Key features:

(i) Short-term vehicle parking (if applicable)

(ii) Commercial kiosks or retail frontage

(iii) Service vehicle access, landscape buffers

Design priority: Moderate—supports broader urban integration and public realm

3.3.4 Additional Considerations

(i) The shape and size of these zones depend on available space, station typology, and street context

(ii) The hierarchy helps prioritize design decisions such as paving materials, lighting levels, signage clarity, and maintenance responsibility

(iii) In limited land, zones may overlap, but pedestrian safety and universal access in primary areas must be maintained

Designing station access areas with a clear zone hierarchy improves user safety, operational efficiency, and spatial legibility—key qualities for high-performing transit environments.

In many projects, access zones were planned too narrowly, leading to conflicts between pedestrians and drop-off vehicles. Clearly mapping zones during early in the design process helps prevent costly changes later.

3.4 Matching Station and Street Typologies

Designing effective transit facilities requires understanding how station types and street contexts interact. The combination of a station's physical form and the character of the surrounding street network strongly influences how space should be allocated, what kind of access facilities are needed, and how the public realm is organized.

This section offers guidance on aligning station typologies with street contexts to inform site planning, design priorities, and access strategies.

Examples of Common Combinations and Their Implications

A. Elevated Station on a Primary Arterial

Challenges: Limited pedestrian safety, high-speed traffic, understructure shadows

Design responses:

(i) Grade-separated pedestrian crossings (bridges or tunnels)

(ii) Use shaded space under viaducts for public seating, bicycle parking, or commercial activity

(iii) Separate pedestrian and vehicular access zones to avoid conflicts

B. Underground Station on a Pedestrian-Priority Street

Advantages: Ample walkability, activated public space, high footfall

Design responses:

(i) Maximize street-level entrances and legibility

(ii) Integrate seating, signage, and greenery to enhance public comfort

(iii) Enable smooth connections to adjacent buildings and shared spaces

(iv) Separate pedestrian and vehicular access zones to avoid conflicts

C. At-Grade Station on a Collector Street

Considerations: Shared road usage, moderate traffic, mixed land use

Design responses:

(i) Widen sidewalks and ensure safe mid-block crossings

(ii) Provide clear PUDO with minimal disruption to traffic

(iii) Allocate curb space for multiple uses (e.g., bike parking, short-term loading)

D. Interchange Terminal in a Mixed-Use Zone

Complexity: High passenger volume, multiple access needs, integration with TOD

Design responses:

(i) Allocate distinct areas for each mode (bus, rail, ride-hailing, walking)

(ii) Ensure universal access through ramps, elevators, and tactile cues

(iii) Coordinate with developers on building interface and shared amenities)

TRANSIT FACILITY ELEMENTS BY MODE 4

Transit stations function as critical hubs where various modes of transport converge. The design quality for the first and last segments of a passenger's journey—on foot, by bicycle, by feeder bus, or through ride-hailing services—significantly impacts the usability, safety, and overall user experience.

Transit facility elements in sidewalks, bus bays, bicycle parking, PUDO zones, and vehicle access points ensure smooth and accessible transitions between modes and define the station area's functional performance. The following sections in this chapter comprise a mode-by-mode guide on the planning and design of transit facility elements located in the unpaid public realm.

Each mode includes a discussion on the following:

(i) key design principles with recommended spatial layouts and features;

(ii) typical facility elements identifying design principles and best practices;

(iii) design considerations for each mode; and

(iv) illustrative tips on safety, accessibility, and operational considerations.

4.1 Pedestrian Access

Pedestrian infrastructure is the foundation of transit access. Every passenger is a pedestrian at some point in their journey—whether approaching from nearby housing, transferring between modes, or navigating the station area itself. A well-designed pedestrian environment ensures safe, intuitive, and universally accessible movement to and from the station.

This section outlines key considerations for designing sidewalks, crossings, and related facilities in the unpaid zone around transit stations.

4.1.1 Key Design Principles

Continuity: Provide uninterrupted sidewalks that connect directly to station entrances, bus stops, and drop-off areas.

Visibility and safety: Ensure clear sightlines at crossings and avoid visual obstructions from parked vehicles or street furniture.

Universal accessibility: All pedestrian paths must be barrier-free and usable by people with disabilities, older people, and children.

Separation from vehicles: Use curbs, plant buffers, or bollards to protect pedestrians from adjacent traffic or drop-off zones.

Comfort and shade: Provide shade, seating, lighting, and adequate width to support high footfall and waiting passengers.

4.1.2 Typical Facility Elements

Guidelines in designing the elements for pedestrian infrastructure are presented in Table 2.

Table 2: Pedestrian Design Guidelines

Element	Design Guidelines
Sidewalk Width	Minimum 2.5–3.0 m in low-traffic areas; 4.0–5.0 m or more in high-volume station zones
Tactile Paving	Use to guide visually impaired users from sidewalks to station entries and crossings
Ramps and Curb Cuts	Slopes no steeper than 1:12, minimum 1.2 m width; avoid abrupt level changes
Crosswalks	Signalized or raised crossings at desire lines; ensure visual contrast and lighting
Buffer Zones	Minimum 0.5–1.0 m setback between sidewalk and traffic or drop-off areas

m = meter.
Source: Asian Development Bank.

4.1.3 Design Considerations for Station Areas

(i) Avoid clutter near entrances (e.g., signage, vendors) that may block pedestrian flow.
(ii) Design crossing points to align with actual pedestrian desire lines, not just street geometry.
(iii) Include seating or rest points every 50–100 m for older people or passengers carrying luggage.
(iv) Ensure sufficient wayfinding to guide people from surrounding neighborhoods and other modes.

4.1.4 Illustrative Tip

Where sidewalks are constrained, prioritize accessible circulation and shaded rest areas. Figure 10 shows how cantilevered or arcaded walkways in dense urban areas can reclaim pedestrian space without widening roads.

Figure 10: Pedestrian Access in Constrained Urban Environments

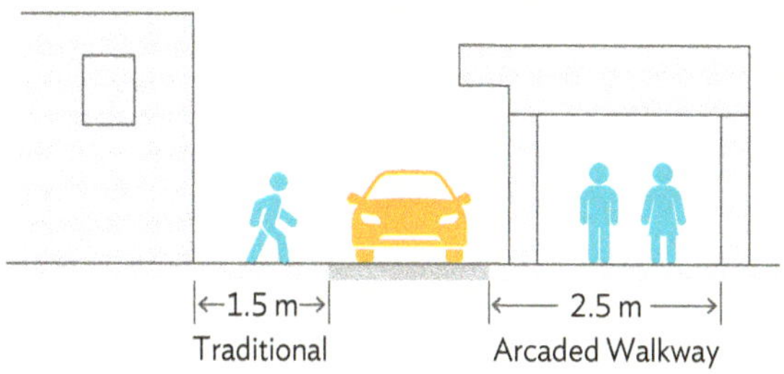

For streets with no wide sidewalks, consider cantilevered or arcaded walkways.

m = meter.
Source: Asian Development Bank.

A station's pedestrian environment is one of the most visible expressions of transit quality.

By prioritizing comfort, safety, and accessibility in pedestrian design, transit systems can become truly inclusive and welcoming for all users.

4.2 Bicycle Access and Parking

Bicycles offer a flexible, low-cost, and sustainable mode of first- and last-mile access to transit stations. With safe infrastructure and convenient facilities, they can greatly extend a station's reach, especially in medium-density urban and suburban areas.

This section outlines key design considerations for safe bicycle access routes and secure parking facilities in the public realm around transit stations.

4.2.1 Key Design Principles

Directness and connectivity: Bicycle access should connect directly to station entrances, feeder routes, and surrounding neighborhoods.

Safety and separation: Provide protected bike lanes or dedicated bike paths where possible; minimize conflicts with pedestrians and vehicles.

Visibility and wayfinding: Bicycle routes and parking must be clearly signed and easy to locate from a distance.

Security and weather protection: Parking should offer protection from theft, vandalism, and weather to encourage regular use.

Integration with other modes: Ensure that bicycle parking is well-located relative to pedestrian paths, bus stops, and drop-off zones.

4.2.2 Typical Facility Elements

Considerations in designing the elements for bicycle access and parking are presented in Table 3.

Table 3: Bicycle Parking Type and Best-Use

Type	Description	Best Use
Open Rack	Simple, low-cost frame for locking front wheel or frame	Low-volume areas or short-term use
Covered Rack	Rack with roof or shelter above	Moderate use areas or weather-sensitive environments
Bike Locker	Fully enclosed, individually assigned unit	Long-term use or high-security locations
Bike Park or Pod	High-capacity, often multitiered system	Terminal or interchange stations

Source: Asian Development Bank.

4.2.3 Design Considerations

(i) For the placement, locate it within 30–50 m of the main station entrance.

(ii) Ensure at least 1.5–2.0 m of maneuvering space around racks for clearance.

(iii) To improve lighting and visibility aspects, avoid placing racks in hidden or poorly lit areas.

(iv) To optimize the capacity, use ridership projections and TOD data to estimate future demand.

(v) Provide accessible bike parking options for cargo bikes, folding bikes, or bikes with child seats to comply with more universal design.

4.2.4 Illustrative Tip

Combine signage, pavement markings, and bollards to clearly define bike entry routes. Where space allows, integrate bicycle parking with seating or planting to create more attractive and multi-use public space (Figure 11).

Figure 11: Integrated Station—Street Context Matching Framework (MRT Jakarta Context)

Source: Asian Development Bank.

Investing in safe, visible cycling infrastructure promotes active mobility, reduces congestion at drop-off points, and fosters healthier, more connected communities.

4.3 Public Transport Integration

Seamless integration between urban rail stations and surface public transport—such as buses, BRT, or informal minibuses—is essential for maximizing network efficiency and improving access for users beyond walking or cycling distance. Poor integration can lead to long transfer times, unsafe walking routes, and passenger confusion.

This section provides a design guide for bus stop placement, transfer zones, and related infrastructure to support effective multimodal interchange near stations.

4.3.1 Key Design Principles

Proximity and directness: Locate bus stops and bays as close to the station entrance as possible—ideally within 30–50 meters.

Safe and accessible transfer: Provide wide, step-free, and protected pedestrian paths between the station and bus boarding points.

Clear separation of flows. Avoid conflicts between pedestrians, buses, and other vehicles by using signage, pavement markings, and barriers.

Weather protection and amenities. Provide shaded or covered waiting areas, benches, lighting, and wayfinding signs to improve comfort.

Operational flexibility. Design curb space or bays to accommodate multiple services and vehicle sizes without congestion or unsafe boarding.

4.3.2 Typical Facility Elements

Considerations in designing the elements for integration facilities are presented in Table 4.

Table 4: Bus Facility Type and Use Case

Facility Type	Description	Use Case
In-Line Bus Stop	Bus stops directly along the main street curb, with minimal station-side infrastructure	Low to moderate demand corridors
Bus Bay or Pull-Out	Recessed bus stop that allows buses to pull out of the traffic lane	Medium to high-traffic streets; allows smoother flow
Dedicated Bus Terminal or Loop	Larger off-street area for terminating routes, layovers, or interchange	High-volume stations or terminal nodes

Source: Asian Development Bank.

4.3.3 Design Considerations for Station Areas

(i) Waiting area width of at least 2.5 m; wider for multiple routes or high frequency

(ii) Boarding area should align with bus floor height for universal access

(iii) Clearly display route numbers, destinations, and headways

(iv) Use turning radii that accommodate bus size without impacting pedestrian space

(v) Work with operators and agencies to ensure that route scheduling and stop placement are aligned with station hours and user demand

4.3.4 Illustrative Tip

Where space is constrained, coordinate multiple feeder services to use a single high frequency loop. Prioritize drop-off over layover to reduce curb demand and passenger congestion.

Public bus integration elements at transit station. (left photo) a multi-level interchange integrating mass rapid transit, bus rapid transit, and feeder buses in close proximity to reduce transfer time; (right photo) a well-organized bus interchange with multiple bays, clear separation of pedestrian and bus flows, and weather-protected waiting areas, supporting efficient multimodal transfers (photos by Jimmytst/Dreamstime.com and justinekessytz).

Better public transport integration at station areas creates a cohesive, user-friendly transit network, reduces car dependence, and improves access equity—especially for outer districts and lower-income communities.

4.4 Ride-Hailing, Taxi, and Private Vehicles

In many urban settings, ride-hailing services (e.g., app-based taxis), conventional taxis, and private vehicles remain common modes for accessing transit stations, especially during early morning, late evening, or in low-density catchment areas. Well-designed PUDO areas help maintain pedestrian safety and smooth traffic flow.

This section outlines how to accommodate ride-hailing, taxi, and private vehicle access to ensure organized, safe, and supportive of multimodal integration.

4.4.1 Key Design Principles

Separation from pedestrian flows: Avoid mixing PUDO zones with main pedestrian access routes to enhance safety.

Efficient turnover: Prioritize short dwell times and enforce time limits to prevent queuing or illegal parking.

Visibility and legibility: Use pavement paint, signage, and lighting to clearly identify pick-up zones.

Accessibility and proximity: Place zones within 30–60 m of the station entrance, with barrier-free paths to access points.

Coordination across modes: Ensure PUDO areas do not block bus bays, bicycle lanes, or station service routes.

4.4.2 Typical Facility Elements

Considerations in designing the elements for ride-hailing, taxi, and private vehicle facilities are presented in Table 5.

Table 5: Pick-Up and Drop-Off Facility Types and Recommended Use Case

Type	Description	Suitable For
On-Street PUDO Bay	Simple curbside zone marked for PUDO only	Low-volume or shared-use roads
Median Pick-up Point	Center-lane pick-up with U-turn access and waiting refuge	High-volume streets with limited curb space
Off-Street Loop or Pocket	Dedicated area set back from main road, shared by taxis and private cars	High-demand or interchange stations

PUDO = pick-up and drop-off.
Source: Asian Development Bank.

4.4.3 Design Considerations

(i) Use curb management tools (e.g., signage, painted curbs, enforcement) to limit waiting

(ii) Add lighting, bollards, and signage to prevent conflicts with pedestrians or cyclists

(iii) Where possible, integrate digital queuing systems or app-based vehicle identification to reduce idling and congestion

(iv) Ensure step-free paths with appropriate turning space for wheelchair-accessible vehicles from PUDO zones to station entrances

4.4.4 Illustrative Tip

In space-constrained locations, use a shared PUDO zone with clear time-of-day allocation (e.g., taxis in morning peak, ride-hailing in the afternoon and evening). Mark each section with large icons and painted zones to ensure smooth passenger flow.

Figure 12: Public Bus Integration Elements at Transit Station

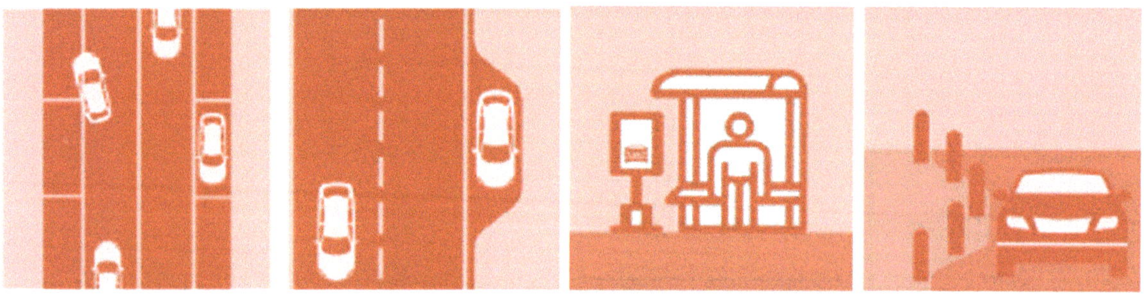

Parking
Normally in the form of off-street parking facility

Layby
Readily accessible from public street

Waiting Area for Passengers
Area with shelter for passengers waiting to be picked up, with dropped kerb for ease of access by wheelchair/assisted walking users

Traffic Calming Features/Bollard
For slowing down of cars entering from the public roads into the ladyby area

Source: Asian Development Bank.

Organized PUDO areas reduce conflicts and congestion, ensuring vehicle access supports rather than disrupts multimodal transit operations.

4.5 Motorcycle and Car Parking

While transit systems aim to reduce private vehicle dependency, motorcycle and car parking remain essential in many station areas—especially where public transport is not the dominant mode or feeder service is limited. Well-planned parking supports modal shift while minimizing impacts on space, safety, and pedestrian movement. This section outlines key design considerations for short-term and long-term parking near stations.

4.5.1 Key Design Principles

Peripheral placement: Locate parking facilities outside of primary pedestrian access zones to reduce congestion near station entrances.

Clear User Segmentation: Separate motorcycle parking from car parking and delineate areas for public, staff, and special-needs users.

Controlled access and security: Provide lighting, surveillance, or attendants, especially for motorcycles, to prevent theft and misuse.

Flexible layouts: Design parking that can adapt over time (e.g., converted to bike parking or green space).

Pricing and policy alignment: Coordinate parking policy (e.g., time limits or pricing) with broader goals to encourage walking, cycling, or feeder use.

4.5.2 Typical Facility Elements

Considerations in designing the elements for motorcycle and car parking facilities are presented in Table 6.

Table 6: Parking Facility Types and Recommended Use Case

Type	Description	Use Case
On-Street Parking	Painted bays or marked curbs	Short-stay or informal access
Surface Lot	Open air paved area for cars or bikes	Suburban or land-rich location
Structured Parking	Multilevel facility, often shared with transit-oriented development or mall	High-demand, high-density areas
Park-and-Ride	Larger facility intended for full-day storage	Outlying stations or terminal stops

Source: Asian Development Bank.

4.5.3 Design Considerations

(i) Motorcycle parking dimensions should be 0.6 m x 2.0 m (minimum) per unit; ensure a stable surface and access paths.

(ii) Car parking dimensions should be 2.5 m x 5.0 m (standard); wider for accessible bays.

(iii) Ensure separation from pedestrians flows with physical buffers or clear markings.

(iv) Indicate time limits, pricing (if any), and user types clearly.

(v) Include accessible parking within 50 m of station entrance, with barrier-free connection.

4.5.4 Illustrative Tip

In high-demand areas, reserve the most convenient spaces for motorcycles, as they require less space and reduce congestion. Design parking zones as modular platforms that can be repurposed for bicycles, ride-hailing, or public space. This future-proofing allows easy reconfiguration as mobility needs evolve, without major infrastructure changes.

5 EVALUATION AND SCORING FRAMEWORK

Designing high-quality transit facilities requires not just applying principles, but measuring how effectively they are implemented. To support consistent quality, transparency, and continuous improvement, this chapter introduces an Evaluation and Scoring Framework to assess the following elements of the public realm design surrounding transit stations:

(i) Pedestrian facilities, such as sidewalks, crossings, ramps, and covered walkways

(ii) Bicycle infrastructure, including access routes and secure parking

(iii) Pick-up and drop-off areas for private vehicles, ride-hailing, and taxis

(iv) Bus stops and feeder vehicle bays that support seamless intermodal transfers

(v) Waiting areas, signage, and wayfinding systems that enhance user orientation and comfort

(vi) Accessibility features, such as tactile paving, curb ramps, and universal design elements

(vii) Public space elements, such as landscaping, lighting, seating, and safety features that enhance the quality and usability of the area around the station

The framework will enable government agencies, planners, designers, and stakeholders to

(i) assess whether station access and connectivity designs meet core user-focused objectives,

(ii) identify areas for improvement early in the planning or review process, and

(iii) track the performance of facility upgrades or network-wide interventions over time.

The framework translates into measurable evaluation criteria the following planning principles for people-centered transit facilities, which will help users understand how well a design performs to reach these core objectives:

(i) enhance multimodal connectivity,

(ii) promote people-centric connectivity,

(iii) foster universal accessibility,

(iv) promote social and public well-being, and

(v) promote sustainability and future-proofing systems.

This evaluation framework will help ensure that investments in transit infrastructure translate into functional, inclusive, and people-oriented spaces, whether used for design, audits, or planning.

5.1 How the Evaluation Framework Works

The evaluation framework is practical, adaptable, and usable for various transit station contexts. It supports early design decision-making, technical review processes, and post-construction audits with a consistent way to assess public realm facilities quality around stations.

Step 1: Identify the Station Typology

Before applying the framework, the evaluator should determine the station typology (Table 7). This affects the expectations for access infrastructure and helps prioritize which scoring criteria should carry more weight.

Table 7: Station Typology

Type	Description
Type I: Major Multimodal Station	High-volume, multimodal hub typically located in central business districts or urban centers, serving as a transfer point between multiple lines or modes
Type II: Regular Multimodal Station	Medium-volume station located along major corridors; usually serves a surrounding neighborhood or activity center with moderate access complexity
Type III: Nonmultimodal Station	Local, low-volume station primarily accessed by pedestrians and informal feeders; often located in residential or mixed-use areas

Source: Asian Development Bank.

The station type determines expected facility level and guides score interpretation.

Step 2: Define the Evaluation Area

The framework focuses exclusively on the unpaid public realm, generally defined as the zone within a radius of approximately 50–400 meters from the station entrance, including

(i) pedestrian facilities, such as sidewalks, crossings, ramps, and covered walkways;

(ii) bicycle infrastructure, including access routes and secure parking;

(iii) pick-up and drop-off areas for private vehicles, ride-hailing, and taxis;

(iv) bus stops and feeder vehicle bays that support seamless intermodal transfers;

(v) waiting areas, signage, and wayfinding systems that enhance user orientation and comfort;

(vi) accessibility features, such as tactile paving, curb ramps, and universal design elements; and

(vii) public space elements, such as landscaping, lighting, seating, and safety features that enhance the quality and usability of the area around the station.

Private building interiors, fare-paid areas, and commercial zones are excluded unless they directly interface with the unpaid realm (e.g., integrated entrances or shared access routes).

Step 3: Conduct the Evaluation

Using the scoring matrix (see Appendix), the evaluator scores each criterion based on presence, quality, and appropriateness. Scores range from 0 (absent or poor) to 3 (well-designed and context-sensitive).

Each evaluation may be conducted by

(i) design consultants (self-assessment),

(ii) government reviewers (design approval),

(iii) project managers (site audits), or

(iv) civil society or community observers (public accountability).

Step 4: Interpret the Results

The cumulative score out of 15 is interpreted against benchmarks that are adjusted based on station typology. This ensures that lower-volume stations are not unfairly penalized for infrastructure that may not be relevant or needed. This scoring is not a certification, but a decision support tool—offering a snapshot of strengths and gaps to guide improvement.

5.2 Evaluation Criteria and Scoring Matrix

A scoring assessment framework is organized around five objectives, each supported by strategies and guidelines to be applied during the planning and design of new or upgraded transit stations. The framework assigns separate scores to each subcomponent using quantifiable metrics. The evaluation process and scoring steps are illustrated in Figures 13–17.

Figure 13: Scoring for Objective 1

Objective 1: Multimodal connectivity enhancement

Provide multimodal network connectivity that will add new dimension to travel choices

Criteria Description

Pedestrian: Share of Safe and Accessible Walkway Segments

Measures the percentage of walkway segments within 400 m of the station that are:

- continuous, safe, and accessible to all users (including people with disabilities);
- free from major obstructions or hazards; and
- equipped with adequate lighting.

How to calculate:

Divide the number of qualifying walkway segments by the total segments in the area.

Scoring

Pedestrian

Percentage of walkway segments that are safe and all-accessible

	Type I – Major multimodal station	Type II – Regular multimodal station	Type III – Nonmultimodal station
More than 50% of the walkway is safe and accessible to all users	2 points	2 points	2 points
10%–50% of the walkway is safe and accessible to all users	1 points	1 points	1 points
0%–10% of the walkway is safe and accessible to all users	0 points	0 points	0 points

continued on next page

Figure 13 *continued*

Criteria Description	Scoring

Bicycle: Share of Safe and Accessible Cycle Lane Segments and Provision of Cycle Parking at Station

A. Measures the percentage of cycle lane segments within 400 m of the station that are:

- physically protected from traffic (for streets >30 kph);
- on low-speed streets (≤30 kph) or shared streets (≤15 kph); and
- designed for safe and all-accessible cycling.

How to calculate:

Divide the number of compliant cycle segments by the total number of segments in the area.

B. Assesses whether the station provides secured, fixed bicycle parking (e.g., multispace racks). A station without this facility is marked as noncompliant.

Bicycle

(i) Percentage of cycle lane segments that are safe and all-accessible

	Type I – Major multimodal station	Type II – Regular multimodal station	Type III – Nonmultimodal station
More than 50% of the cycle lane is safe and accessible to all users	2 points	2 points	2 points
10%–50% of the cycle lane is safe and accessible to all users	1 points	1 points	1 points
0%–10% of the cycle lane is safe and accessible to all users	0 points	0 points	0 points

(ii) Provision of cycle parking at station

	Type I – Major multimodal station	Type II – Regular multimodal station	Type III – Nonmultimodal station
Yes	1 point	1 point	1 point
No	0 point	0 point	0 point

Bus: Transit Facility with MRT (Bus–MRT Interchange Proximity)

Measures how many bus stops are available for interchange with the MRT and how far they are from the station entrance.

How to calculate:

For each MRT station, count the number of bus stops nearby and measure their distance from the station. A "stop" refers to a designated boarding point for any bus service.

Bus

(i) No public bus stops interchange with MRT station

	Type I – Major multimodal station	Type II – Regular multimodal station	Type III – Nonmultimodal station
For Type I Station: More than 6 interchanging stops with MRT station	3 points	-	-
For Type I Station: Between 3–5 interchanging stops with MRT station *For Type II/III Station:* Yes	2 points	2 points	2 points
For Type I Station: Between 1–2 interchanging stops with MRT station *For Type II/III Station:* No	1 points	0 points	0 points
For Type I Station: None	0 points	-	-

(ii) Distance between interchange stops (average distance of all stops)

	Type I – Major multimodal station	Type II – Regular multimodal station	Type III – Nonmultimodal station
Within 0–200 m	2 points	2 points	2 points
Within 200–400 m	1 points	1 points	1 points
>400 m	0 points	0 points	0 points

continued on next page

Figure 13 *continued*

Criteria Description	Scoring

Minibus: Transit Facility with MRT (Minibus Interchange Proximity)

Measures the number of minibus (*angkot*) stops available for transfer to MRT and their distance from the station.

How to calculate:

For each MRT station, count nearby minibus stops and measure their distance to the station entrance.

Minibus (*Angkot*)

(i) Provision of interchange facility with MRT station

	Type I – Major multimodal station	Type II – Regular multimodal station	Type III – Nonmultimodal station
Yes	2 points	2 points	2 points
No	0 points	0 points	0 points

(ii) Distance between interchange stops (average distance of all stops)

	Type I – Major multimodal station	Type II – Regular multimodal station	Type III – Nonmultimodal station
Within 0–200 m	2 points	2 points	2 points
Within 200–400 m	1 points	1 points	1 points
>400 m	0 points	0 points	0 points

Light Rapid Transit: Transit Facility with MRT (LRT Interchange Proximity)

Measures the number of LRT and MRT interchange stations and the distance between the two station entrances.

How to calculate:

For each MRT station, identify nearby LRT stations that function as interchanges and measure their distance from the MRT entrance.

LRT

(i) Provision of interchange facility with MRT station

	Type I – Major multimodal station	Type II – Regular multimodal station	Type III – Nonmultimodal station
Yes	3 points	3 points	-
No	0 points	0 points	-

(ii) Distance between interchange stops (average distance of all stops)

	Type I – Major multimodal station	Type II – Regular multimodal station	Type III – Nonmultimodal station
Within 0–200 m	2 points	2 points	-
Within 200–400 m	1 points	1 points	-
>400 m	0 points	0 points	-

continued on next page

Figure 13 *continued*

Criteria Description	Scoring			
Regional Rail: Transit Facility with MRT (Regional Rail Interchange Proximity) Measures the number of interchanges between MRT and Regional Rail (e.g., KRL Commuter Line or Airport Express) and their distance from the MRT station. **How to calculate:** For each MRT station, identify nearby regional rail stations functioning as interchanges and measure the distance between their entrances.	**Regional Rail (Commuter and/or Airport Line)**			
	(i) Provision of interchange facility with MRT station	Type I – Major multimodal station	Type II – Regular multimodal station	Type III – Nonmultimodal station
	Yes	3 points	3 points	-
	No	0 points	0 points	-
	(ii) Distance between interchange stops (average distance of all stops)			
	Within 0–200 m	2 points	2 points	-
	Within 200–400 m	1 points	1 points	-
	>400 m	0 points	0 points	-
MRT: Transit Facility with MRT (MRT–MRT Interchange Proximity) Measures the number of interchange stations between different MRT lines and their distance from each other. **How to calculate:** For each MRT station, identify nearby MRT stations on other lines that serve as interchanges, and measure the distance between their entrances.	**MRT**			
	(i) Provision of interchange facility with MRT station	Type I – Major multimodal station	Type II – Regular multimodal station	Type III – Nonmultimodal station
	Yes	3 points	3 points	-
	No	0 points	0 points	-
	(ii) Distance between interchange stops (average distance of all stops)			
	Within 0–200 m	2 points	2 points	-
	Within 200–400 m	1 points	1 points	-
	>400 m	0 points	0 points	-

continued on next page

Figure 13 *continued*

Criteria Description	Scoring			
Car Park Assesses whether the station provides dedicated car-related facilities, such as: ■ park-and-ride areas; ■ pick-up/drop-off (kiss-and-ride) zones; and ■ designated bays for staff and operations. **How to calculate:** For each station, identify the presence of these car facilities.	**Car**			
	(i) Provision of dedicated bays for car parks			
		Type I – Major multimodal station	Type II – Regular multimodal station	Type III – Nonmultimodal station
	Yes	1 points	1 points	1 points
	No	0 points	0 points	0 points
	(ii) Provision of dedicated bays for kiss-n-ride			
	Yes	2 points	2 points	2 points
	No	0 points	0 points	0 points
Taxi and/or E-hailing Services Assesses whether the station provides dedicated facilities for taxi and e-hailing services, such as: ■ parking or waiting areas with circulation space; and ■ on-street pick-up/drop-off with proper taper lane. **How to calculate:** For each station, identify the presence of these taxi or e-hailing service facilities.	**Taxi**			
	(i) Provision of dedicated bays for taxi parks			
		Type I – Major multimodal station	Type II – Regular multimodal station	Type III – Nonmultimodal station
	Yes	1 points	1 points	1 points
	No	0 points	0 points	0 points
	(ii) Provision of dedicated bays for pick-up/drop-off			
	Yes	2 points	2 points	2 points
	No	0 points	0 points	0 points

continued on next page

Figure 13 *continued*

Criteria Description	Scoring
E-hailing Motorcycle Assesses whether the station provides facilities for motorcycle-based e-hailing services, such as: ▪ dedicated parking or waiting areas with internal circulation; and ▪ on-street stopping with proper taper lane. **How to calculate:** Check if these facilities are available at each transit station.	E-hailing Motorcycle (i) Provision of dedicated bays for e-hailing motorcycle parking Type I – Major multimodal station / Type II – Regular multimodal station / Type III – Nonmultimodal station Yes ▶ 1 points / 1 points / 1 points No ▶ 0 points / 0 points / 0 points
Motorcycle Parking Assesses whether the station provides dedicated motorcycle parking facilities, such as: ▪ parking areas with adequate circulation space; and ▪ on-street access with proper taper lanes. **How to calculate:** For each station, check for the presence of the above motorcycle parking facilities.	Motorcycle (i) Provision of dedicated bays for motorcycle parking Type I – Major multimodal station / Type II – Regular multimodal station / Type III – Nonmultimodal station Yes ▶ 1 points / 1 points / 1 points No ▶ 0 points / 0 points / 0 points

kph = kilometers per hour, LRT = light rapid transit, m = meter, MRT = mass rapid transit.
Source: Asian Development Bank.

Scoring Summary for Objective 1

Objective 1: Multimodal connectivity enhancement

Provide multimodal network connectivity that will add new dimension to travel choices

Type	Max. Score
Type I – Major Multimodal Station	37 points
Type II – Regular Multimodal Station	36 points
Type III – Nonmultimodal Station	21 points

Figure 14: Scoring for Objective 2

Objective 2: Promote people-centric connectivity

Make first-mile last-mile travel more convenient and efficient

Criteria Description	Scoring

Wayfinding:

Assesses whether the station has wayfinding tools to help users navigate, such as:

- maps of nearby destinations and interchanges;
- entrance naming or numbering;
- directional signage within and around the station; and
- tourist information maps (if applicable).

How to calculate:

Check if the station includes the wayfinding elements listed above.

Wayfinding

	Type I – Major multimodal station	Type II – Regular multimodal station	Type III – Nonmultimodal station
(i) Provision of major destination map around station			
Yes	2 points	2 points	2 points
No	0 points	0 points	0 points
(ii) Provision of interchange map at major interchanging stations			
Yes	1 points	1 points	-
No	0 points	0 points	-
(iii) Entrance naming strategy and system			
Yes	2 points	2 points	2 points
No	0 points	0 points	0 points
(iv) Provision of directional signage in and around station			
Yes	2 points	2 points	2 points
No	0 points	0 points	0 points
(v) Provision of information on tourist places surrounding the station			
Yes	1 points	-	-
No	0 points	-	-

continued on next page

Figure 14 *continued*

Criteria Description	Scoring

Pedestrian:

Assesses whether the station provides pedestrian connections to:

- nearby destinations (e.g., offices, malls, parks, schools);
- other public transport stops (e.g., bus, LRT, MRT, *angkot*); and
- street furniture for comfort (e.g., benches, lighting, bins).

How to calculate:

Check if the station includes these pedestrian linkages and amenities.

Pedestrian

(i) Pedestrian connection to nearest major developments and public transit

	Type I – Major multimodal station	Type II – Regular multimodal station	Type III – Nonmultimodal station
Distance between 50–350 m	3 points	3 points	3 points
Distance between 350–500 m	2 points	2 points	2 points
>500 m	1 points	1 points	1 points

(ii) Provision of the pedestrian facility–public furniture (streetlamps, benches, bins)

	Type I – Major multimodal station	Type II – Regular multimodal station	Type III – Nonmultimodal station
5 or more	1 points	1 points	1 points
Less than 5	0 points	0 points	0 points

At-grade crossing for pedestrian or cyclist:

Assesses the provision of complete, safe, and accessible pedestrian or cyclist crossings at intersections, especially on roads with speeds >30 kph.

Qualifying features include:

- barrier-free design for people with disabilities;
- minimum 2 m width and marked clearly;
- refuge islands for wide roads; and
- speed bumps 5–10 m before crossing.

How to calculate:

Count all required intersections, then count how many meet the above criteria. Divide to get the percentage of complete crossings.

Pedestrian and/or Cycling Lane Crossing

(i) Percentage of walkway/cycleway segments with safe-all accessible crossing

	Type I – Major multimodal station	Type II – Regular multimodal station	Type III – Nonmultimodal station
More than 50%	2 points	2 points	2 points
10%–50%	1 points	1 points	1 points
0%–10%	0 points	0 points	0 points

Integration of weather cover:

Assesses the percentage of walkway segments that provide adequate weather protection, such as:

- covered walkways, arcades, or shelters;
- shade from trees or nearby buildings; and
- climate-appropriate features (e.g., air-conditioned paths, if applicable).

How to calculate:

Count all walkway segments, then count how many have effective shade or shelter. Divide to get the percentage of protected walkways.

Integration of weather cover

(i) Integration of walkway segments with all weather protector and/or covered

	Type I – Major multimodal station	Type II – Regular multimodal station	Type III – Nonmultimodal station
More than 50%	2 points	2 points	2 points
10%–50%	1 points	1 points	1 points
0%–10%	0 points	0 points	0 points

continued on next page

Figure 14 *continued*

Criteria Description	Scoring			
Vertical or elevation connection: Assesses the provision of elevators and escalators at the station based on: station type (I, II, III); passenger volume; and interchange needs. **How to calculate:** Identify the station type, then count the number of elevators and escalators provided.	**Vertical or Elevation Connection** – Number of lifts and/or escalators provided for seamless access	Type I – Major multimodal station	Type II – Regular multimodal station	Type III – Nonmultimodal station
	More than 8	3 points	-	-
	For Type I Station: Between 5 to 8; *For Type II/III Station:* More than 4	2 points	2 points	2 points
	Between 1 to 4	1 points	1 points	1 points
	None	0 points	0 points	0 points
Transit-related information dissemination: Assesses whether the station provides passenger information and assistance, including: ticket counters and vending machines; visual, audio, and braille information; real-time service info (e.g., seat availability); and staff or customer assistance for emergencies. **How to calculate:** Check if the station includes the information facilities listed above.	**Information Center**	Type I – Major multimodal station	Type II – Regular multimodal station	Type III – Nonmultimodal station
	(i) Provision of ticket counter			
	Yes	2 points	2 points	2 points
	No	0 points	0 points	0 points
	(ii) Provision of real-time information on seat availability			
	Yes	1 points	-	-
	No	0 points	-	-
	(iii) Provision of customer assistance			
	Yes	1 points	1 points	1 points
	No	0 points	0 points	0 points

kph = kilometers per hour, LRT = light rapid transit, m = meter, MRT = mass rapid transit.

Source: Asian Development Bank.

Scoring Summary for Objective 2

Objective 2: Promote people-centric connectivity

Make first-mile last-mile travel more convenient and efficient

Type	Max. Score
Type I – Major Multimodal Station	24 points
Type II – Regular Multimodal Station	21 points
Type III – Nonmultimodal Station	20 points

Figure 15: Scoring for Objective 3

Objective 3: Foster universal accessibility

Design an ideal transit-built environment that connects with users' senses, including touch, sight, hearing, and ability to remember in support of their use of the system

Criteria Description	Scoring			
Access to and in the station: Assesses whether the station provides universal accessibility through: ■ wheelchair-accessible elevators and ramps; and ■ wide ticket gates that support wheelchair users. **How to calculate:** Check if the station includes the access features listed above.	Access to station			
	(i) Provision of 2-way escalators	Type I – Major multimodal station	Type II – Regular multimodal station	Type III – Nonmultimodal station
	2-way escalators	2 points	2 points	2 points
	Only 1-way escalators	1 points	1 points	1 points
	No provision of escalators	0 points	0 points	0 points
	(ii) Provision of wheelchair-accessible elevator			
	Yes	2 points	2 points	2 points
	No	0 points	0 points	0 points
	(iii) Provision of wide ticket gates for wheelchair access			
	Yes	2 points	2 points	2 points
	No	0 points	0 points	0 points
Nursery room: Assesses whether the station provides a nursery room equipped with baby changing and feeding facilities. **How to calculate:** Check if the station includes a nursery room with the features described above.	Nursery Room			
	(i) Provision of nursery room for users with babies	Type I – Major multimodal station	Type II – Regular multimodal station	Type III – Nonmultimodal station
	Yes	1 points	1 points	1 points
	No	0 points	0 points	0 points

continued on next page

Figure 15 *continued*

Criteria Description	Scoring

Barrier-free access facility:

Assesses whether the station is equipped for universal access, including:

- tactile paths and wayfinding maps for visually impaired users; and
- information provided in visual, audio, and braille formats.

How to calculate:

Check if the station includes all the barrier-free facilities listed above.

Barrier-free Facilities

(i) Provision of tactile path to assist visually impaired users

	Type I – Major multimodal station	Type II – Regular multimodal station	Type III – Nonmultimodal station
Yes	2 points	2 points	2 points
No	0 points	0 points	0 points

(ii) Provision of a tactile wayfinding guide to assist visually impaired users

	Type I – Major multimodal station	Type II – Regular multimodal station	Type III – Nonmultimodal station
Yes	1 points	1 points	1 points
No	0 points	0 points	0 points

(iii) Provision of an information by visual, sound, and braille

	Type I – Major multimodal station	Type II – Regular multimodal station	Type III – Nonmultimodal station
Yes	1 points	1 points	1 points
No	0 points	0 points	0 points

Seating areas:

Assesses whether the station provides seating in key areas, including:

- platform;
- paid area;
- unpaid area; and
- with both regular and priority seating.

How to calculate:

Check if seating is available in all required areas and count the number of seats provided.

Seating Areas

(i) Number of seating

	Type I – Major multimodal station	Type II – Regular multimodal station	Type III – Nonmultimodal station
More than 10 seats	2 points	2 points	2 points
Between 1–10 seats	1 points	1 points	1 points
None	0 points	0 points	0 points

(ii) Number of benches provided at the paid station area

	Type I – Major multimodal station	Type II – Regular multimodal station	Type III – Nonmultimodal station
Yes	1 points	–	–
No	0 points	–	–

(iii) Number of benches provided at un-paid area

	Type I – Major multimodal station	Type II – Regular multimodal station	Type III – Nonmultimodal station
Yes	1 points	–	–
No	0 points	–	–

continued on next page

Figure 15 *continued*

Criteria Description	Scoring			
Child-friendly design: Assesses whether the station includes child-friendly features such as: - low-height sinks; and - safety barriers at key locations. **How to calculate:** Check if the station includes the child-friendly elements described above.	**Child-friendly Design** (i) Provision of child-friendly design (railing, sink height, etc.)	Type I – Major multimodal station	Type II – Regular multimodal station	Type III – Nonmultimodal station
	Yes	1 points	1 points	1 points
	No	0 points	0 points	0 points

kph = kilometers per hour, LRT = light rapid transit, m = meter, MRT = mass rapid transit.
Source: Asian Development Bank.

Scoring Summary for Objective 3

Objective 3: Foster universal accessibility

Design an ideal transit-built environment that connects with users' senses, including touch, sight, hearing and ability to remember in support of their use of the system

Type	Max. Score
Type I – Major Multimodal Station	16 points
Type II – Regular Multimodal Station	14 points
Type III – Nonmultimodal Station	14 points

Figure 16: Scoring for Objective 4

Objective 4: Promote social and public well-being

Enrich community engagement by creating a safer, cared-for environment at the station

Criteria Description	Scoring

Social Activity:

Assesses whether the station includes spaces that encourage community interaction, such as:

- art or heritage exhibition areas;
- food and vending stalls;
- retail kiosks or delivery lockers; and
- advertising space.

How to calculate:

Check if the station provides any of the above social activity facilities.

Social Activity

	Type I – Major multimodal station	Type II – Regular multimodal station	Type III – Nonmultimodal station
(i) Provision of reserve area for food, art, and heritage exhibition (for both short term and long term)			
Yes	2 points	2 points	2 points
No	0 points	0 points	0 points
(ii) Provision of commercial shops and/or e-commerce delivery booth			
Yes	3 points	3 points	3 points
No	0 points	0 points	0 points
(iii) Provision of advertising space			
Yes	2 points	2 points	2 points
No	0 points	0 points	0 points

Entertainment activity:

Assesses whether the station provides space for entertainment activities such as:

- music events;
- street performances; and
- other public events.

How to calculate:

Check if the station includes a designated area for entertainment or event activities.

Entertainment Activity

Provision of reserve area for entertainment events

	Type I – Major multimodal station	Type II – Regular multimodal station	Type III – Nonmultimodal station
Yes	1 points	-	-
No	0 points	-	-

continued on next page

Figure 16 *continued*

Criteria Description	Scoring

Educational and information:

Assesses whether the station provides space for educational or informational purposes, such as:

- kiosks for book collection or learning activities; and
- community news and activity boards.

How to calculate:

Check if the station includes any of the educational or community information facilities described above.

Educational and Information Activity

(i) Provision of reserve area for educational awareness programs

	Type I – Major multimodal station	Type II – Regular multimodal station	Type III – Nonmultimodal station
Yes	1 points	–	–
No	0 points	–	–

(ii) Provision of local community news and activities board

	Type I – Major multimodal station	Type II – Regular multimodal station	Type III – Nonmultimodal station
Yes	1 points	1 points	1 points
No	0 points	0 points	0 points

(iii) Provision of information related to crime deterrence and/or prevention

	Type I – Major multimodal station	Type II – Regular multimodal station	Type III – Nonmultimodal station
Yes	1 points	1 points	1 points
No	0 points	0 points	0 points

Safety:

Assesses whether the station provides essential safety features, including:

- emergency assembly point;
- CCTV and crime prevention systems;
- smart lighting with emergency call functions; and
- fire extinguishers, emergency intercoms, and alert systems.

How to calculate:

Check if all required safety facilities are present. Stations without adequate safety provisions must be flagged, regardless of overall score.

Safety

(i) Emergency Meeting Point

	Type I – Major multimodal station	Type II – Regular multimodal station	Type III – Nonmultimodal station
Yes	3 points	3 points	3 points
No	0 points	0 points	0 points

(ii) Provision of surveillance measures and fast crime reporting installations

	Type I – Major multimodal station	Type II – Regular multimodal station	Type III – Nonmultimodal station
Yes	3 points	3 points	3 points
No	0 points	0 points	0 points

Source: Asian Development Bank.

Scoring Summary for Objective 4

Objective 4: Promote social and public well-being

Enrich community engagement by creating a safer, cared-for environment at the station

Type	Max. Score
Type I – Major Multimodal Station	17 points
Type II – Regular Multimodal Station	15 points
Type III – Nonmultimodal Station	15 points

Figure 17: Scoring for Objective 5

Objective 5: Instil sustainability and future-proofing design

Encourage adaptable designs for all transit facilities that require minimum cost in the future to resize and/or redevelop

Criteria Description	Scoring				
Integration of greenery/landscape into connection/station surroundings: Assesses whether the station includes green design features such as: ▪ landscaped walkways; ▪ tree-lined paths; and ▪ vertical planting or green walls. **How to calculate:** Check if the station area includes any of the greenery or landscape features described.	Integration of Greenery Percentage area allocated for green area in the station		Type I – Major multimodal station	Type II – Regular multimodal station	Type III – Nonmultimodal station
	Yes	▶	1 points	-	-
	No	▶	0 points	-	-
Robust and future-proof design and sizing: Assesses whether the station is designed to support future expansion and emerging mobility needs, including: ▪ reserved space for future escalators, lifts, or reconfiguration; and ▪ EV charging points or docking stations for e-scooters and/or e-bikes. **How to calculate:** Check if the station includes structural provisions or facilities that support future growth and technology integration.	Future-proof Design (i) Provision of reserve area at platform and concourse for future additions such as stairs/lift/ramps		Type I – Major multimodal station	Type II – Regular multimodal station	Type III – Nonmultimodal station
	Yes	▶	2 points	2 points	2 points
	No	▶	0 points	0 points	0 points
	(i) Provision/reserve for future installation of charging for EVs, charging/docking for e-scooters and/or e-bicycles				
	Yes	▶	2 points	2 points	2 points
	No	▶	0 points	0 points	0 points

continued on next page

Figure 17 *continued*

Criteria Description	Scoring			
Future development plan: Assesses whether the station design includes safeguards for future TOD, such as: - structural provisions; - reserved connectivity links; and - space for parking or PUDO integration. **How to calculate:** Check if the station incorporates design elements that align with planned or proposed TOD developments.	Future Development plan Provision of transit-oriented development design safeguards	Type I – Major multimodal station	Type II – Regular multimodal station	Type III – Nonmultimodal station
	Yes	2 points	2 points	2 points
	No	0 points	0 points	0 points

Source: Asian Development Bank.

Scoring Summary for Objective 5

Objective 5: Instil sustainability and future-proofing systems

Encourage adaptable designs for all transit facilities that require minimum cost in the future to resize and/or redevelop

Type	Max. Score
Type I – Major Multimodal Station	7 points
Type II – Regular Multimodal Station	6 points
Type III – Nonmultimodal Station	6 points

Below is a sample summary of a scoring assessment. The detailed scoring assessment framework is attached in the Appendix.

Objective	Description	Type I – Major Multimodal Station	Type II – Regular Multimodal Station	Type III – Nonmultimodal Station
1	Multimodal connectivity	37	36	21
2	People-centric connectivity	24	21	20
3	Universal accessibility	15	13	13
4	Social and public wellbeing	17	15	15
5	Sustainability and future proofing systems	7	6	6
Grand Total		**101**	**92**	**76**

Each station is evaluated using a point-based framework across several subcomponents. Scores reflect how well the station meets design goals based on its typology:

(i) Type I: ≥ 61 points

(ii) Type II: ≥ 55 points

(iii) Type III: ≥ 46 points

Stations scoring below the 60% threshold may receive recommendations from authorities and are expected to improve facilities within a set timeline.

APPENDIX: DETAILED SCORING MATRICES AND ASSESSMENT

Objective: Multimodal connectivity

Elements	Metrics	Type I – Major Multimodal Station		Type II – Regular Multimodal Station		Type III – Nonmultimodal Station	
		Criteria	Points	Criteria	Points	Criteria	Points
Pedestrian	Percentage of walkway segments that are safe and all-accessible	More than 50%	2	More than 50%	2	More than 50%	2
		10%–50%	1	10%–50%	1	10%–50%	1
		0%–10%	0	0%–10%	0	0%–10%	0
Bicycle	Percentage of cycle lane segments that are safe and all-accessible	More than 50%	2	More than 50%	2	More than 50%	2
		10%–50%	1	10%–50%	1	10%–50%	1
		0%–10%	0	0%–10%	0	0%–10%	0
	Provision of cycle parking at station	Yes	1	Yes	1	Yes	1
		No	0	No	0	No	0
Bus	Number of public bus stops interchange with MRT stations	More than 6 stops	3	Yes	2	Yes	2
		3–5 stops	2				
		1–2 stops	1	No	0	No	0
		None	0				
	Distance between interchange stops (average distance of all stops)	Within 0–200 m	2	Within 0–200 m	2	Within 0–200 m	2
		200–400 m	1	200–400 m	1	200–400 m	1
		>400 m	0	>400 m	0	>400 m	0
Angkot	Number of *Angkot* stops interchange with MRT stations	Yes	2	Yes	2	Yes	2
		No	0	No	0	No	0
	Distance between interchange stops (average distance of all stops)	Within 0–200 m	2	Within 0–200 m	2	Within 0–200 m	2
		200–400 m	1	200–400 m	1	200–400 m	1
		>400 m	0	>400 m	0	>400 m	0

continued on next page

Continued

Objective: Multimodal connectivity

Elements	Metrics	Type I – Major Multimodal Station		Type II – Regular Multimodal Station		Type III – Nonmultimodal Station	
		Criteria	Points	Criteria	Points	Criteria	Points
LRT	Number of LRT stops interchange with MRT stations	Yes	3	Yes	3		
		No	0	No	0		
	Distance between interchange stops	Within 0–200 m	2	Within 0–200 m	2		
		200–400 m	1	200–400 m	1		
		>400 m	0	>400 m	0		
Regional Rail	Number of regional rail (commuter and/or airport express) stops interchange with MRT stations	Yes	3	Yes	3		
		No	0	No	0		
	Distance between interchange stops	Within 0–200 m	2	Within 0–200 m	2		
		200–400 m	1	200–400 m	1		
		>400 m	0	>400 m	0		
MRT	Number of MRT stops interchange with MRT stations	Yes	3	Yes	3		
		No	0	No	0		
	Distance between interchange stops	Within 0–200 m	2	Within 0–200 m	2		
		200–400 m	1	200–400 m	1		
		>400 m	0	>400 m	0		
Car	Provision of dedicated bays for car parks	Yes	1	Yes	1	Yes	1
		No	0	No	0	No	0
	Provision of dedicated bays for kiss-n-ride	Yes	2	Yes	2	Yes	2
		No	0	No	0	No	0
Taxi and/or E-hailing	Provision of dedicated bays for taxi parks	Yes	1	Yes	1	Yes	1
		No	0	No	0	No	0
	Provision of dedicated bays for kiss-n-ride	Yes	2	Yes	2	Yes	2
		No	0	No	0	No	0
E-hailing motorcycle	Provision of pooling area for e-hailing motorcycle	Yes	1	Yes	1	Yes	1
		No	0	No	0	No	0
Motorcycle parking	Provision of dedicated bays for motorcycle parking	Yes	1	Yes	1	Yes	1
		No	0	No	0	No	0
Subtotal	**Maximum score**	**37**		**36**		**21**	

continued on next page

Continued

Objective: Localized people-centric connectivity

Elements	Metrics	Type I – Major Multimodal Station		Type II – Regular Multimodal Station		Type III – Nonmultimodal Station	
		Criteria	Points	Criteria	Points	Criteria	Points
Wayfinding	Provision of major destination map around station	Yes	2	Yes	2	Yes	2
		No	0	No	0	No	0
	Provision of interchange map at major interchanges	Yes	1	Yes	1		
		No	0	No	0		
	Entrance naming strategy and system	Yes	2	Yes	2	Yes	2
		No	0	No	0	No	0
	Provision of directional signage in and around station (within 400 m radius)	Yes	2	Yes	2	Yes	2
		No	0	No	0	No	0
	Provision of information on a tourist places surrounding the station	Yes	1				
		No	0				
Pedestrian	Pedestrian connection to nearest major developments and public transit	Distance between 50–350 m	3	Distance between 50–350 m	3	Distance between 50–350 m	3
		Distance between 350–500 m	2	Distance between 350–500 m	2	Distance between 350–500 m	2
		Distance > 500 m	1	Distance > 500 m	1	Distance > 500 m	1
	Pedestrian facilities – public furniture (street lamp, bench, bins, etc.)	5 or more	1	5 or more	1	5 or more	1
		Less than 5	0	Less than 5	0	Less than 5	0
At-grade crossing	Percentage of walkway and/or cycleway segments with safe, all-accessible pedestrian/cycle crossing	More than 50%	2	More than 50%	2	More than 50%	2
		10%–50%	1	10%–50%	1	10%–50%	1
		0%–10%	0	0%–10%	0	0%–10%	0
Integration of weather cover	Percentage of walkway segments with all-weather protector and/or covered	More than 50%	2	More than 50%	2	More than 50%	2
		10%–50%	1	10%–50%	1	10%–50%	1
		0%–10%	0	0%–10%	0	0%–10%	0

continued on next page

Continued

Objective: Localized people-centric connectivity

Elements	Metrics	Type I – Major Multimodal Station		Type II – Regular Multimodal Station		Type III – Nonmultimodal Station	
		Criteria	Points	Criteria	Points	Criteria	Points
Vertical and/or elevation on connections	Number of lifts and/or escalators provided for seamless access	More than 8	3	More than 4	2	More than 4	2
		Between 5 and 8	2				
		Between 1 and 4	1	Between 1 and 4	1	Between 1 and 4	1
		None	0	None	0	None	0
Information Centre	Provision of ticket counter	Yes	2	Yes	2	Yes	2
		No	0	No	0	No	0
	Provision of real time information of seat availability	Yes	1				
		No	0				
	Provision of customer assistance	Yes	1	Yes	1	Yes	1
		No	0	No	0	No	0
Subtotal	**Maximum score**	**24**		**21**		**20**	

Objective: Universal accessibility

Elements	Metrics	Type I Criteria	Points	Type II Criteria	Points	Type III Criteria	Points
Access to station	Provision of 2-way escalators, elevators, ramp wheelchair access	2-way escalators	2	2-way escalators	2	2-way escalators	2
		Only 1-way escalators	1	Only 1-way escalators	1	Only 1-way escalators	1
		No	0	No	0	No	0
	Provision of wheelchair-accessible elevator	Yes	2	Yes	2	Yes	2
		No	0	No	0	No	0
	Provision of wide ticket gates for wheel chair access	Yes	2	Yes	2	Yes	2
		No	0	No	0	No	0
Nursery rooms	Provision of nursery room for users with babies	Yes	1	Yes	1	Yes	1
		No	0	No	0	No	0
Barrier free facilities	Provision of tactile path to assist visually impaired users	Yes	2	Yes	2	Yes	2
		No	0	No	0	No	0
	Provision of a tactile wayfinding guide to assist visually impaired users	Yes	1	Yes	1	Yes	1
		No	0	No	0	No	0
	Provision of information by visual, sound, and braille	Yes	1	Yes	1	Yes	1
		No	0	No	0	No	0

continued on next page

Continued

Objective: Universal accessibility

Elements	Metrics	Type I – Major Multimodal Station		Type II – Regular Multimodal Station		Type III – Nonmultimodal Station	
		Criteria	Points	Criteria	Points	Criteria	Points
Seating areas	Number of benches provided at platform area	More than 10 seats	2	More than 10 seats	2	More than 10 seats	2
		Between 1 and 10 seats	1	Between 1 and 10 seats	1	Between 1 and 10 seats	1
		None	0	None	0	None	0
	Number of benches provided at paid station area	Yes	1				
		None	0				
	Number of benches provided at un-paid area	Yes	1				
		None	0				
Child Friendly Design	Provision of child friendly design (railing, sink height, etc.)	Yes	1	Yes	1	Yes	1
		No	0	No	0	No	0
Subtotal	**Maximum score**	**16**		**14**		**14**	

Objective: Social and public wellbeing

Elements	Metrics	Criteria	Points	Criteria	Points	Criteria	Points
Social Activity	Provision of reserve area for food, art and heritage exhibition (for both short term and long term)	Yes	2	Yes	2	Yes	2
		No	0	No	0	No	0
	Provision of commercial shops and/or e-commerce delivery booth	Yes	3	Yes	3	Yes	3
		No	0	No	0	No	0
	Provision of advertising space	Yes	2	Yes	2	Yes	2
		No	0	No	0	No	0
Entertainment Activity	Provision of reserve area for entertainment events	Yes	1				
		No	0				

continued on next page

Continued

Objective: Social and public wellbeing

Elements	Metrics	Type I – Major Multimodal Station		Type II – Regular Multimodal Station		Type III – Nonmultimodal Station	
		Criteria	Points	Criteria	Points	Criteria	Points
Educational and Information	Provision of reserve area for educational and/or awareness programs	Yes	1				
		No	0				
	Provision of local community news and activities board	Yes	1	Yes	1	Yes	1
		No	0	No	0	No	0
	Provision of information related to crime deterrence and/or prevention	Yes	1	Yes	1	Yes	1
		No	0	No	0	No	0
Safety	Emergency Meeting Point	Yes	3	Yes	3	Yes	3
		No	0	No	0	No	0
	Provision of surveillance measures and fast crime reporting installations	Yes	3	Yes	3	Yes	3
		No	0	No	0	No	0
Subtotal	**Maximum score**	**17**		**15**		**15**	
Objective: Sustainability and future proofing systems							
Integration of greenery	Percentage of area allocated for green area in the station	Yes	1				
		No	0				
Future-proof Design	Provision of reserve area at platform and concourse for future additions such as stairs/lift/ramps	Yes	2	Yes	2	Yes	2
		No	0	No	0	No	0
	Provision and/or reserve for future installation of charging for EVs, charging and/or docking for e-scooters and/or e-bicycles	Yes	2	Yes	2	Yes	2
		No	0	No	0	No	0
Future Development plan	Provision of transit oriented development design safeguards	Yes	2	Yes	2	Yes	2
		No	0	No	0	No	0
Subtotal	**Maximum score**	**7**		**6**		**6**	
Grand Total	**Maximum score**	**101**		**92**		**76**	

kph = kilometers per hour, LRT = light rapid transit, m = meter, MRT = mass rapid transit.

Source: Asian Development Bank.

www.ingramcontent.com/pod-product-compliance
Lightning Source LLC
Chambersburg PA
CBHW041758220926
51847CB00014B/250
* 9 7 8 9 2 9 2 7 7 7 5 8 6 *